LOW-ENERGY FPGAs – ARCHITECTURE AND DESIGN

THE KLUWER INTERNATIONAL SERIES IN ENGINEERING AND COMPUTER SCIENCE

LOW-ENERGY FPGAs – ARCHITECTURE AND DESIGN

by

Varghese George

STMicroelectronics

and

Jan M. Rabaey

University of California, Berkeley

SPRINGER SCIENCE+BUSINESS MEDIA, LLC

ISBN 978-1-4613-5545-8 ISBN 978-1-4615-1421-3 (eBook)
DOI 10.1007/978-1-4615-1421-3

Library of Congress Cataloging-in-Publication Data

A C.I.P. Catalogue record for this book is available
from the Library of Congress.

Printed on acid-free paper.

To our friends and family

TABLE OF CONTENTS

PREFACE

This books deals with the energy consumption of Field-Programmable Gate Arrays (FPGAs). Field-Programmable Gate Arrays are becoming popular as embedded components in computing platforms. The programmability of the FPGA can be used to customize implementation of functions on an application basis. This can lead to performance gains, and enable reuse of expensive silicon real estate.

To exploit the FPGA completely, the drawback of poor energy efficiency has to be addressed. This work analyzes the energy components of commercial FPGA architectures, and the trend into future process technologies. Based on this, modifications to the FPGA architecture and circuit techniques are evaluated to improve the energy efficiency.

This book is based on the research work by the authors at the University of California at Berkeley as part of the reconfigurable computing project, which required the design of a low-energy FPGA. The different aspects of the energy efficiency of the FPGA are analyzed in the course of designing a low-energy FPGA, LP_PGAII. The presentation of the material follows the design steps culminating in the final implementation.

The material presented will be of interest to designers working in the field of FPGAs. It will also be useful to circuit designers to illustrate how specific characteristics of the architecture can be exploited to improve the energy efficiency.

The book assumes basic understanding of an FPGA and digital circuit design. **Chapter 1** will briefly touch on some of these subjects to bring the reader up to speed. This chapter also introduces the new paradigm of reconfigurable computing that has been made possible by FPGAs.

Chapter 2 looks at the implication of deep-submicron technology on FPGA power dissipation. The commercial FPGA architecture will be evaluated to isolate the effect of the different architectural components on the total energy consumption. This will help in identifying blocks that have to be redesigned to improve the energy efficiency of the FPGA architecture.

The logic and routing resources of an FPGA are limited. Increasing the logic and routing capacity by providing more resources usually increases the area, delay, and energy overhead. Hence, an environment is required to evaluate the effect of architectural modifications on the performance costs. **Chapter 3** describes the exploration environment to guide and evaluate design decisions. The flow covers all of the steps required for implementing an application described as a logic netlist on the target FPGA architecture.

Chapter 4 looks at the architectural optimization process to evaluate the trade-offs between the flexibility of the architecture, and the effect on the performance metrics. It is also an effort to evaluate smarter ways of distributing the routing switches to improve the performance metrics without sacrificing flexibility.

An exhaustive evaluation of the entire architecture design space would be difficult. In this work, a more practical approach is taken, using an existing architecture as a starting point. The Symmetric Mesh architecture that has been researched extensively and is the basic structure in the popular XC4000 series from Xilinx is used as the starting point. The logic block structure and the interconnect architecture are modified to minimize the total energy in the FPGA. The software environment described in Chapter 3 is used to implement the benchmark suite on the target architecture, and extract the performance data.

The optimization at the architectural level must be adequately supported at the circuit level. **Chapter 5** looks at different circuit techniques to reduce the performance overhead of some of the dominant components. Distinctive features of the FPGA environment have to be taken into account while evaluating the feasibility of existing circuit techniques.

Different low-swing signaling techniques are evaluated to determine their feasibility in an FPGA environment. The effect of the sizing of the routing switches on the delay and energy of the interconnect architecture is explored. Characteristics of the clock distribution that are specific to the FPGA environment are exploited to reduce the clocking energy.

One of the dominant application domains envisioned for the FPGA is as a performance accelerator. This involves frequent reconfiguration as the FPGA is used to implement different tasks. The delay and energy overhead due to the configuration become important in such an environment. **Chapter 6** looks at methods to configure the FPGA to minimize the programming overhead.

A physical implementation is invaluable to verify the different low-energy methods. **Chapter 7** looks at the physical realization of some of the critical components and the final implementation of a low-energy FPGA, LP_PGAII.

In **Chapter 8**, the prototype array is compared to an equivalent commercial architecture. To do this comparison, different applications are implemented on the architectures, and performance data is measured.

The authors wish to acknowledge the Electrical Engineering and Computer Sciences department at the University of California at Berkeley, where the research was conducted. We would like to express our thanks to DARPA for funding this project under the Adaptive Computing Systems project. The prototype chips could not have been made without the fabrication facilities and assistance provided by STMicroelectronics. In particular, we are grateful to Bhusan Gupta of the Central Research and Development Laboratory of STMicroelectronics.

We thank the other members in the Pleiades group who contributed considerably during the course of this project. These include Marlene Wan, Hui Zhang, Vandana Prabhu, and Eric Kusse.

The authors acknowledge the discussions with many people in the field. In particular we wish to thank Andre DeHon, John Wawrzynek, and Kurt Keutzer from UC Berkeley. We would also like to thank Michael J. Alexander and Gabriel Robins at University of Virginia for the help in developing the FPGA routing tools.

We wish to acknowledge the help of Jennifer Evans and Anne Murray, of Kluwer Academic Publishers, for the all the help rendered in getting this book out in time.

Varghese George would like to thank Jen Achtert, who has been a constant support throughout this time.

Varghese George
Jan M. Rabaey
Berkeley, CA.

Chapter 1

INTRODUCTION

1 INTRODUCTION

Integrated circuits in the form of Application Specific Integrated Circuits (ASICs) and General-Purpose (GP) processors traditionally have been used to meet the computational needs for processing information. ASICs can be used to realize fixed applications that have to be executed with the minimum amount of area, delay and energy costs. As the size of the application becomes larger, it becomes practically impossible to implement it in silicon. This is where the general-purpose processor steps in. By breaking the application into smaller functions, it is possible to execute each function sequentially. This makes it possible to reuse a single piece of silicon for a variety of tasks. Applications using processors based on a set of general-purpose instructions often results in inefficient implementation. One solution is to enhance the instruction set to provide specialized complex instructions, with dedicated functional units. As the problem space expands, this method will also run out of steam.

The reconfigurable computing domain is aimed at this problem space between ASICs and general-purpose processors. The Field-Programmable Gate Array (FPGA) can be programmed to compute the problem at hand in a spatial fashion. The goal of reconfigurable architectures is to achieve implementation efficiency approaching that of specialized logic while providing the silicon reusability of general-purpose processors. Field-Programmable Gate Array belongs to this class of architectures and analysis of this approach has shown higher functional density than general-purpose processors [DeHon00].

This chapter will introduce some of the basic FPGA concepts. The basic idea behind using the FPGA as a performance accelerator will be introduced, and some of the research projects in the area will be briefly discussed.

2 FPGA

The FPGA can be visualized as programmable logic blocks embedded in programmable interconnect, as shown in Fig. 1. Unlike ASICs, the logic and interconnect resources are uncommitted, and can be configured to implement different logic functions and connectivity.

The functional complexity of logic blocks can vary from simple two-input Boolean operations to larger, complex, multi-bit arithmetic operations. The choice of the logic block granularity is dependent on the target application domain. The interconnect architecture provides the connectivity between logic blocks and is often the bottleneck in FPGA structures.

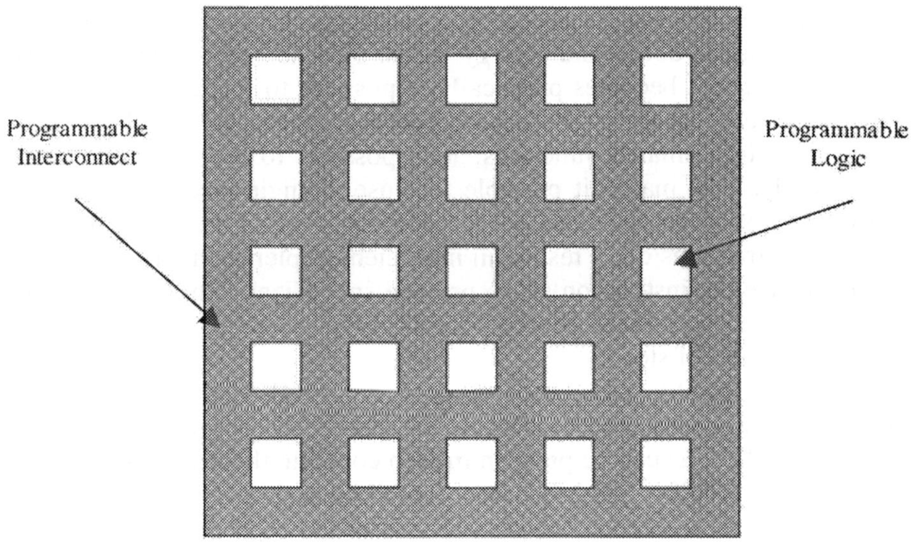

Figure 1. Field-Programmable Gate Array

The programming technology determines the method of storing the configuration information, and comes in different flavors. It has a strong impact on the area and performance of the array. The main programming technologies are: Static Random Access Memory (SRAM), antifuse, and non-volatile technologies using floating gates. The choice of the programming technology is based on the computation environment in which the FPGA is used.

3 INTERCONNECT ARCHITECTURE

The interconnect architecture is realized using switches that can be programmed to realize different connections. The method of providing the connectivity between the logic blocks has a strong impact on the characteristics of the FPGA architecture. The arrangement of the logic and interconnect resources can be broadly classified into four groups: Island style, row-based, sea-of-gates, and hierarchical.

3.1 Island Style Architecture

The island style architecture consists of an array of programmable logic blocks with vertical and horizontal programmable routing channels. The basic architecture is illustrated in Fig. 2.

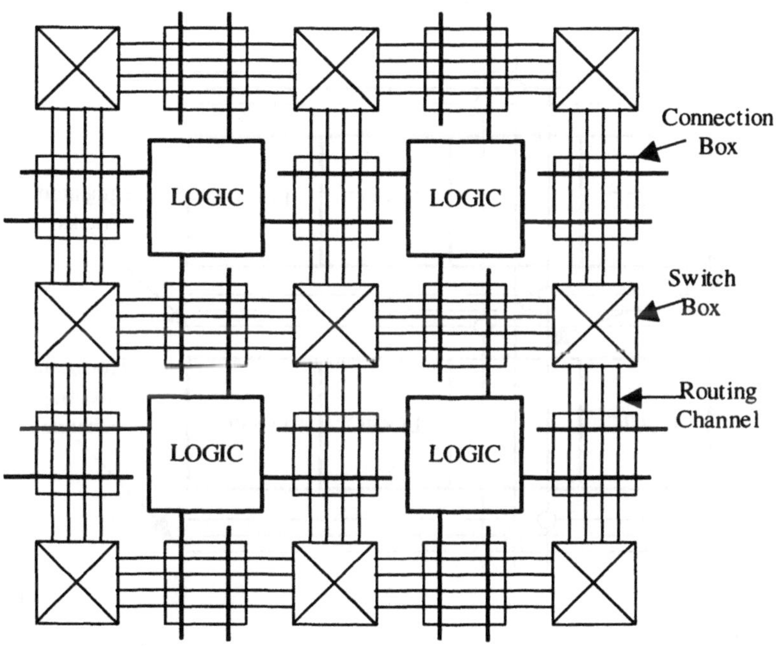

Figure 2. Island Style Architecture

The number of segments in the channel determines the resources available for routing. This is quantified in terms of the channel width, W. The pins of the logic block can access the routing channel through the connection box. The connectivity of each pin to the segments in the channel is determined by the flexibility, F_c, of the connection box. The vertical and horizontal routing channels are connected at the switch box. The flexibility, F_s, of the switch box determines the connections available from each track in a routing channel to the tracks in the other routing channels.

The XC4000 and XC3000 series from Xilinx [Xilinx2] are examples of this kind of architecture.

3.2 Row-Based Architecture

As the name implies, this architecture has logic blocks arranged in rows with horizontal routing channel between successive rows. The row-based architecture is shown in Fig. 3.

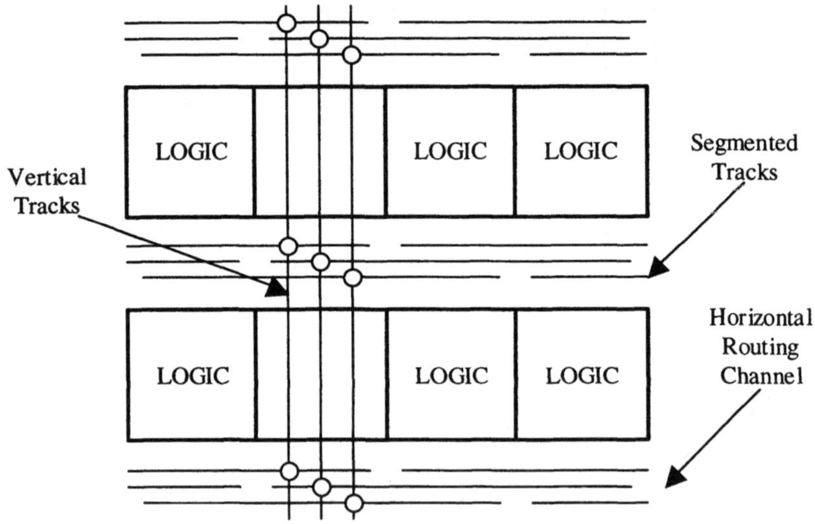

Figure 3. Row-Based Architecture

The routing tracks within the channel are divided into one or more segments. The length of the segments can vary from the width of a module pair to the full length of the channel. The segments can be connected together at the ends using programmable switches to increase their length. Other tracks run vertically through the logic blocks. They provide connections between the horizontal routing channels. The pins of the logic blocks can connect to the horizontal routing channel and the vertical routing segments. The length of the wiring segments in the channel is determined by tradeoffs involving the number of tracks, the resistance of the routing switches, and the capacitance of the segments.

The ACT3 family of FPGAs from Actel [Actel2] is an example of this architecture.

3.3 Sea-of-Gates Architecture

The sea-of-gates architecture, unlike the previous architectures, is not an array of logic blocks embedded in a general routing structure. The architecture consists of fine-grain logic blocks covering the entire floor of the device. Connectivity is realized using dedicated neighbor-to-neighbor routes that are usually faster than general routing resources. Usually the architecture also uses some general routes to realize longer connections. The SX family of FPGAs from Actel [Actel3] is an example of this class of architecture.

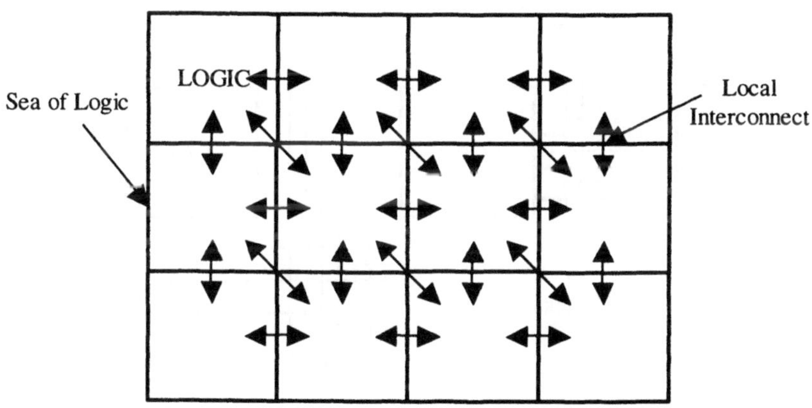

Figure 4. Sea-of-Gates Architecture

The Triptych [Borriello95] FPGA architecture has taken this concept further by removing the distinction between logic and routing resources. The logic blocks are replaced by structures that can perform both logic and routing tasks. This permits a smooth trade-off between logic and routing resource usage.

3.4 Hierarchical Architecture

Most logic designs exhibit locality of connections, which implies a hierarchy in the placement and routing of the connections between the logic blocks. The hierarchical FPGA architecture tries to exploit this feature to provide smaller routing delays and a more predictable timing behavior. This architecture is created by connecting logic blocks into clusters. These clusters are recursively connected to form a hierarchical structure. Fig. 5 illustrates a possible architecture.

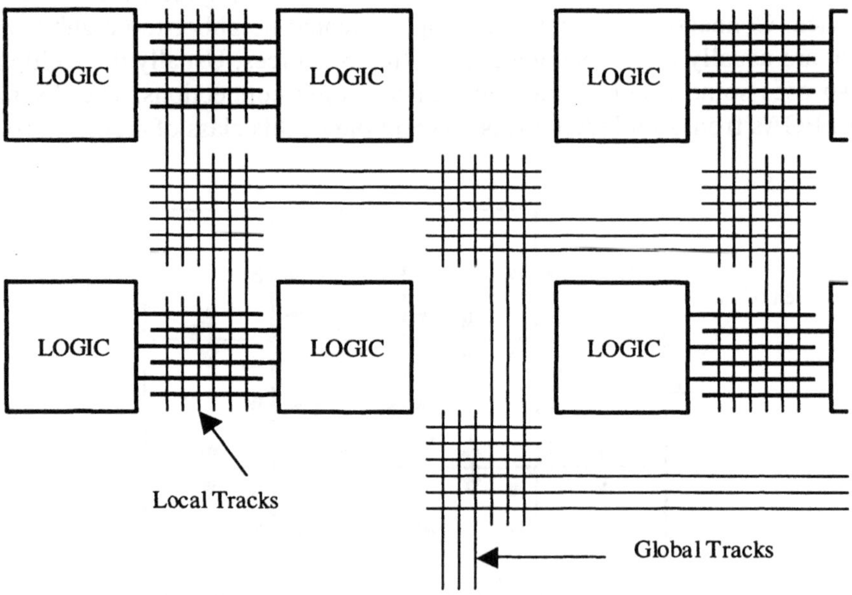

Figure 5. Hierarchical FPGA Architecture

The speed of a net is determined by the number of routing switches it has to pass through. The hierarchical structure reduces the number of switches in series for long connections and can hence potentially run at a higher speed.

The HSRA [Tsu99] and the HFPGA [Lai98] proposed by Yen-Tai, et al. belongs to this class of architectures. The HSRA uses a binary tree structure with pipelined interconnect to achieve high data throughput. Experimental results from the HFPGA recommends a 4-ary tree to obtain area improvements by reducing the total switch count in the architecture.

4 LOGIC BLOCK ARCHITECTURE

The logic block is responsible for implementing the gate level functionality required for each application. The functionality of the logic block can be defined by the number of different functions it can implement. The functionality of the logic block has a direct impact on the routing resources. As the functional capability of the logic block increases, the amount of logic that can be packed into it increases. This reduces the amount of external routing resources. As the logic block size increases, it is also quite possible that the block cannot be fully utilized, resulting in wastage. Based on this tradeoff, there are numerous logic block structures trying to optimize the area and speed of the FPGA.

The functionality of the logic blocks is obtained by controlling the connectivity of some basic logic gates or by using lookup-tables (LUTs). Fig. 6(a) shows the logic block diagram of the macro cell used in Altera FPGAs [Altera3]. Different combinational logic functions are obtained by manipulating the connections in the programmable AND array. The product terms are fed to an OR gate. The result of the combinatorial operation can be registered.

Another approach to the implementation of the programmable logic block is illustrated in Fig. 6(b). This is the block diagram of the Configurable Logic Block (CLB), used in the XC4000 architecture [Xilinx2]. The functionality of the block is determined by the contents of the LUT. This logic block can implement any five-input logic function and selected functions of up to nine variables. It is also possible to implement unrelated smaller functions in a single logic block.

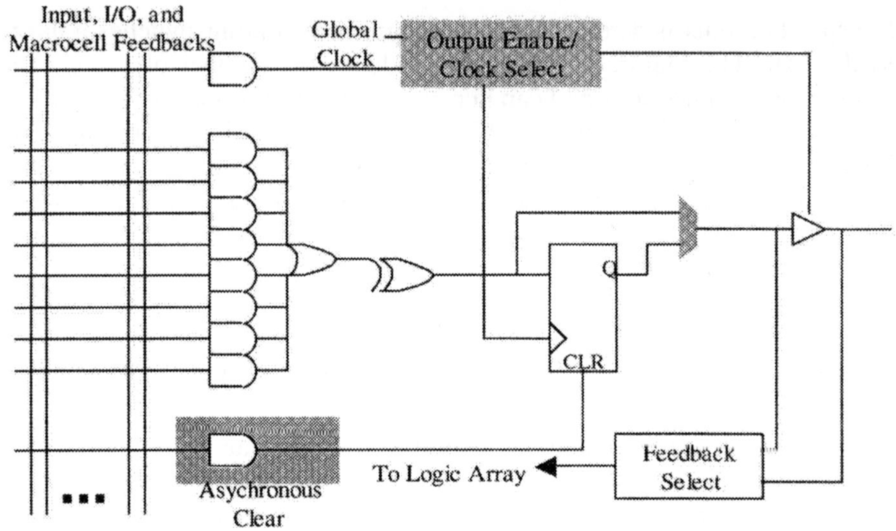

(a) Altera Macro Cell [Altera3]

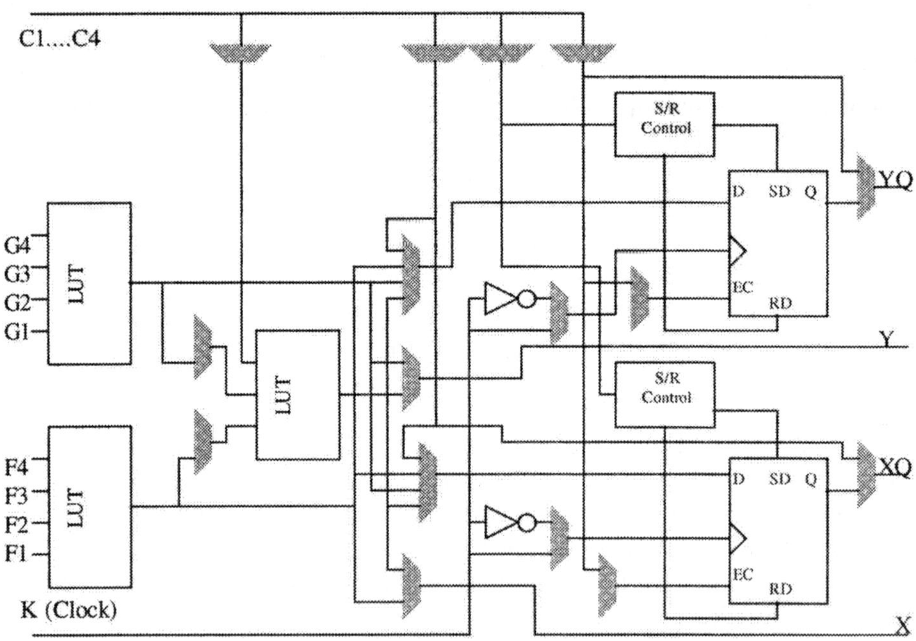

(a) XC4000 CLB [Xilinx]

Figure 6. Logic Block Structures

There is also a wide range in the size of the logic block. The logic block used in the ACT1 architecture [Actel1] represents the finer grain of the spectrum. The MATRIX, with each logic block comparable to an 8-bit ALU, [Mirsky96] represents the coarser granularity. This choice in the logic block granularity is also influenced by the application space envisioned for these FPGAs. The fine-grain programmability is more amenable to control functions while the coarser grain blocks with arithmetic capability are more useful for datapath operations.

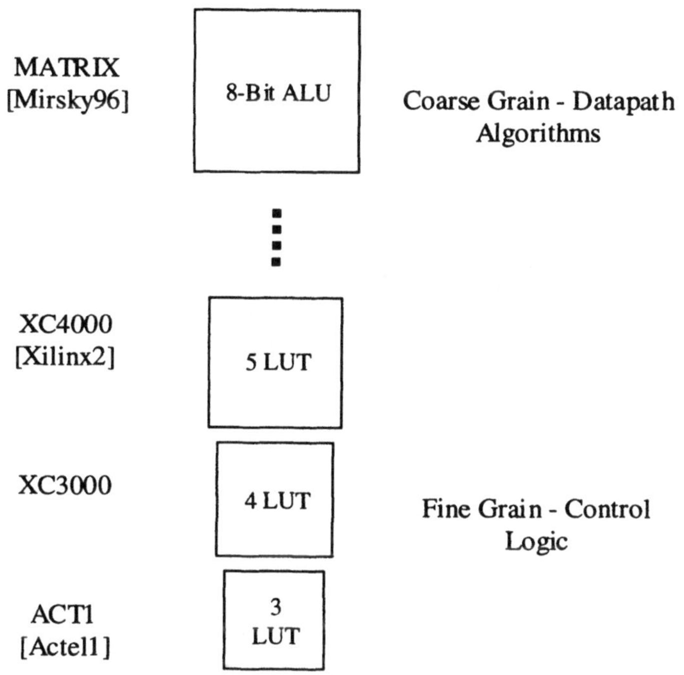

Figure 7. FPGAs and Logic Block Granularity

5 PROGRAMMING TECHNOLOGY

The logic and routing resources of the FPGA are uncommitted, and must be programmed to realize the required behavior. The contents of the logic block can be programmed to control the functionality of the logic block, while the routing switches can be programmed to control the connections between the logic blocks.

There are different methods to store this program information, ranging from the volatile SRAM method to the irreversible antifuse technology. The area of an FPGA is dominated by the area of the programmable components. Hence, the choice of the programming technology can also affect the area of the FPGA. Another factor that has to be considered is the number of times the FPGA has to be programmed. The antifuse-based FPGA can be programmed only once, while the SRAM-based FPGA does not limit the number of times the array can be reprogrammed.

5.1 SRAM

In this method of programming, the configuration is stored in SRAM cells. Fig. 8 shows a five-transistor memory cell used in FPGAs to store the configuration. When the interconnect is implemented using pass-transistors, the SRAM cells control whether the transistor is on or off. In the case of the lookup tables used in the logic block, the logic is stored in the SRAM cells.

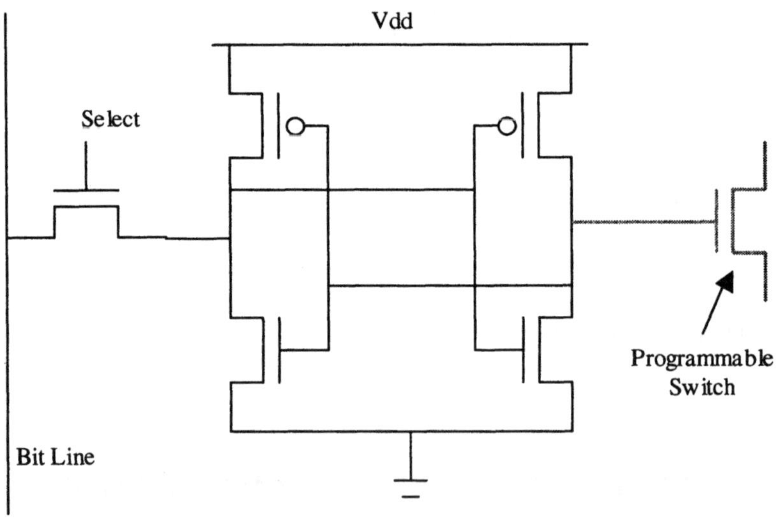

Figure 8. Five-Transistor Configuration Memory Cell

This method does suffer from the fact that the storage is volatile and the configuration has to be written into the FPGA each time on power-up. For systems using SRAM-based FPGAs, an external permanent storage device is usually used.

This technology requires at least five transistors per cell. Due to the relatively large size of the memory cells, the area of the FPGA is dominated by configuration storage.

This method offers the convenience of reusing a single device for implementing different applications by loading different configurations. This characteristic has made SRAM-based FPGAs popular in reconfigurable platforms, which strive to obtain performance gains by customizing the implementation of functions to the specific application.

5.2 Antifuse

In the SRAM programming method, the information is stored by controlling the state of the memory cell. The antifuse programming method uses a programmable connection whose impedance changes on the application of a high voltage. In the un-programmed state, the impedance of the connection is of the order of a few giga-ohms, and can be treated as an open circuit. By applying a high voltage, a physical change called fusing occurs. This results in an impedance of a few ohms though the device, establishing a connection.

The Programmable Low Impedance Circuit (PLICE) [Hamdy88] used in Actel devices is an example of an antifuse. Fig. 9 shows the cross-section of a PLICE antifuse. An antifuse consists of a dielectric sandwiched between a polysilicon layer and a diffusion layer. In the normal state, the dielectric creates high impedance between the two layers. Application of a high voltage results in melting of the dielectric and creation of a connection between the two terminals.

This method has the advantage that the area of the programming element is of the order of the size of a Via, and therefore can achieve a significant reduction in area compared to the SRAM-programmed FPGA. The resistance through the element is of the order of a few ohms and is much smaller than the resistance of a pass-transistor that is used as the routing switch in the SRAM method.

This method is non-volatile, and does not require external configuration storage on power-down. Unlike the SRAM based technology, errors in the design cannot be corrected, since the programming process is irreversible.

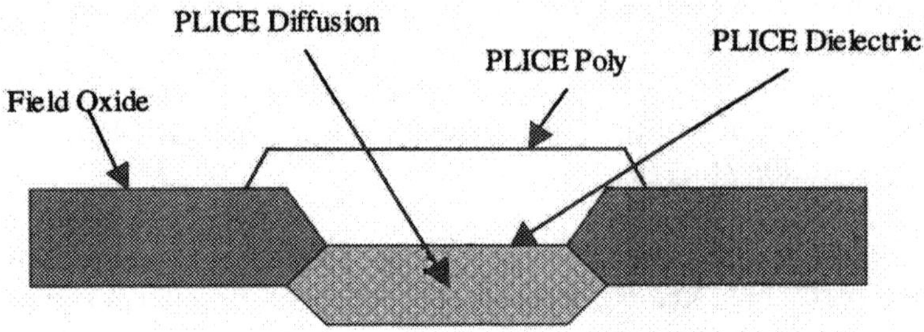

Figure 9. **PLICE Cross Section**

5.3 EPROM, EEPROM, and Flash

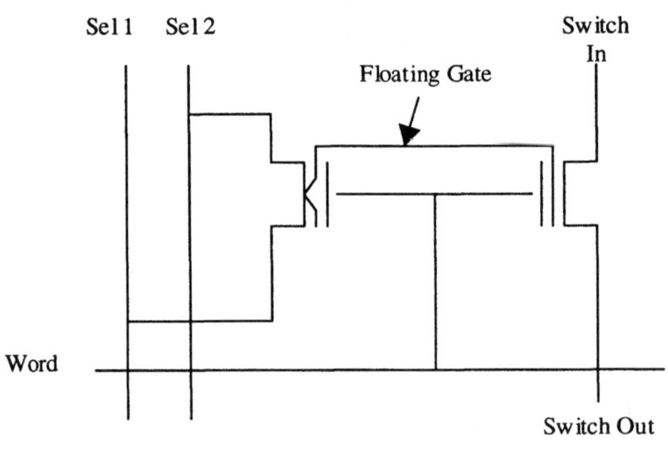

Figure 10. Floating Gate [Actel4]

This class of non-volatile programming technology uses the same techniques as EPROM, EEPROM, and Flash memory technologies. This method uses a special transistor with two gates: a floating gate and a select gate. When a large current flows through the transistor, a charge is trapped in the floating gate that increases the threshold voltage of the transistor. Under normal operation, the programmed transistors will act as open circuits, while the other transistors can be controlled using the select gates. The charge under the floating gate will persist during power-down. The floating charge can be removed by exposing the gate to ultraviolet light in the case of EPROMs, and by electrical means in the case of EEPROMs and Flash.

These techniques straddle the middle ground between the SRAM and antifuse techniques. They provide the non-volatility of antifuse with the reprogrammability of SRAM. The resistance of the routing switches is larger than that of antifuse, while the programming is more complex and time consuming than that of the SRAM technique. The ProASIC family of FPGAs from Actel [Actel4] uses the Flash programming technology.

6 COMPUTATION MODEL

The specialized custom logic, the general-purpose processor, and the FPGA operate in different regions. This can be explained better by looking at the way an application is implemented on the different platforms.

Fig. 11 illustrates how an application made up of sub-functions is realized using an ASIC, FPGA, and a general-purpose processor. In an ASIC, silicon is dedicated to each specific function. In a general-purpose processor, each function is computed sequentially on the same piece of silicon. The memory is used as an intermediate storage for the results of each compute cycle. In the FPGA, the logic is programmed to compute each function. The flow of data between the functions is controlled using the programmable interconnect.

The FPGA is similar to the ASIC in the sense that the computation is done spatially, and can be tailored to the specific application. It is also similar to the general-purpose processor in the aspect of silicon reusability.

In an FPGA, the control of the logic and the interconnect is distributed over the array. The fine grain programmability of the FPGA resources results in higher programming overhead per instruction as compared to a general-purpose processor. If the required functionality does not change, the FPGA does not have to be reprogrammed. The general-purpose processor incurs the instruction overhead on each clock cycle.

The domain of each of the platforms can be described as:

- **ASIC** – Fixed logic.
- **FPGA** – Implementation tailored to application, with low temporal variation of required computations.
- **General-purpose Processor** – Implementation is broken down in terms of available instruction-set, with high temporal variation in required computations.

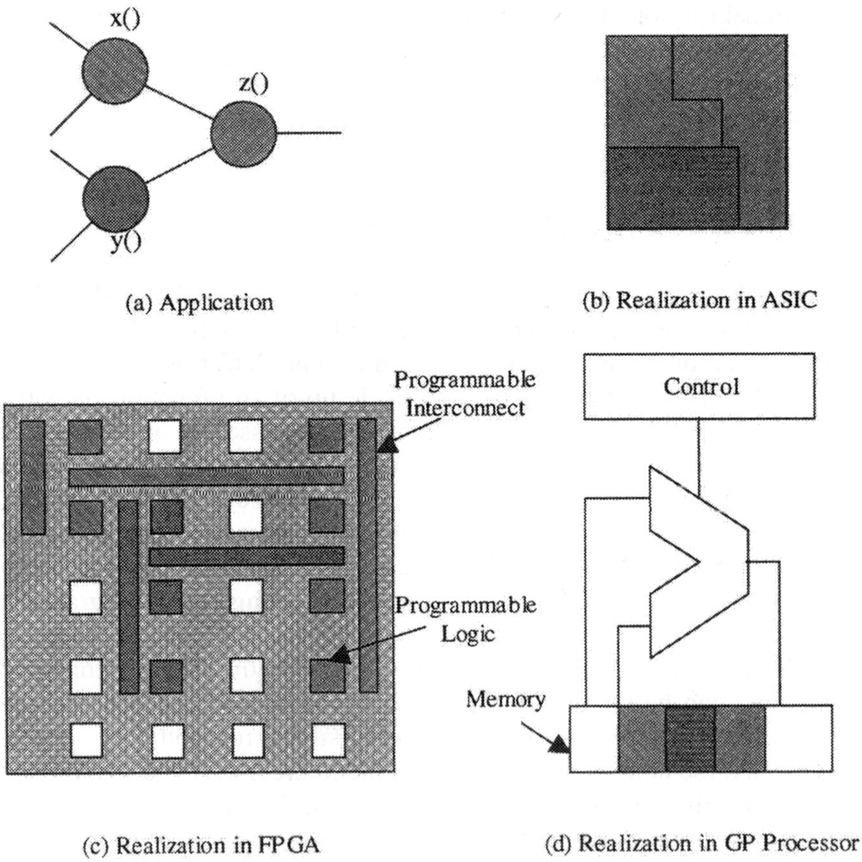

Figure 11. Implementation of a Function in the Different Domains

7 FPGA AS A PERFORMANCE ACCELERATOR

An FPGA can provide the silicon reusability of a general-purpose processor with performance approaching that of an ASIC. This section describes the general principle of the field of reconfigurable computing that takes advantage of the programmability of FPGAs to achieve performance acceleration.

Fig. 12 shows the flow of computation of a sample application. It is made up of functions f_1- f_8. There are two compute intensive loops: g_1 and g_2. Functions f_2, f_3, and f_4 make up g_1. Functions f_6 and f_7 make up g_2. This application can be implemented using an ASIC, GP processor, GP processor with dedicated functional units, or GP processor with FPGA.

The implementations using an ASIC will provide the best performance. However, this involves complex design effort, and dedicated silicon. This will be expensive, especially if the application keeps changing.

This application can also implemented using a GP processor. The functions have to be implemented using the instructions available in the processor. This can often lead to performance degradation. The compute-intensive loops, g_1 and g_2 exhibit temporal locality. Unfortunately, the cycle-by-cycle operation of the processor cannot take advantage of this fact.

The implementation of the compute-intensive loops can be accelerated by using dedicated functional units in conjunction with the GP processor. This assumes the knowledge of the functions g_1 and g_2 at fabrication time, which is not always possible. At the time of implementation of g_2, g_1 is no longer required. This results in the waste of silicon area.

The option explored in the reconfigurable computing paradigm is a coupling of the GP processor and the FPGA. The compute intensive functions can be implemented in the FPGA. The overhead of programming the FPGA places restrictions on the functions that can be efficiently implemented.

In the example, the function f_1 is executed only once. It is preferable to implement this using the processor. However, for the functions g_1 and g_2, the configuration overhead is spread over a number of iterations, and can be tolerated. Since the different functions are realized at run-time, area is not wasted by dedicated silicon implementations.

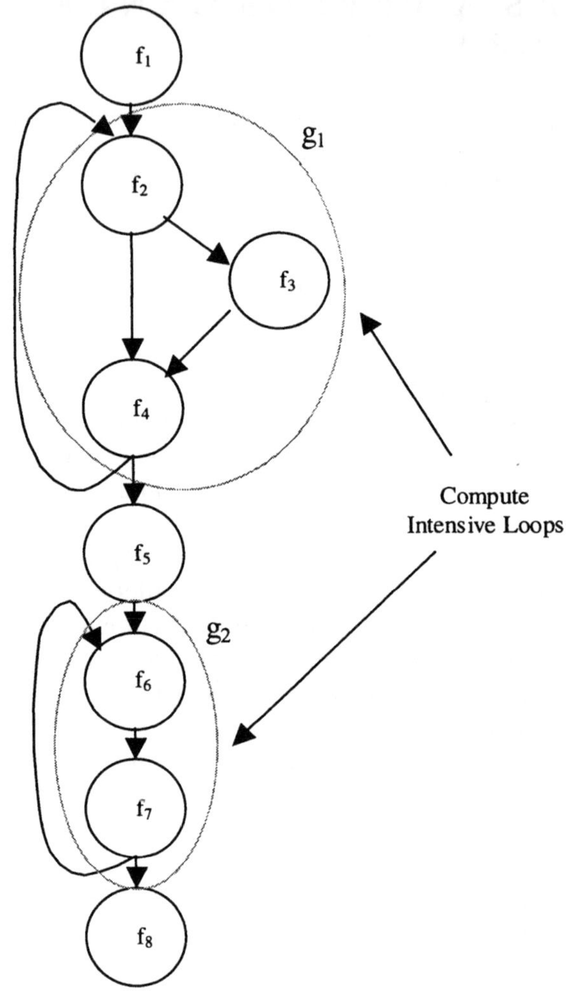

Figure 12. Flow of Computation and Opportunities for Acceleration.

8 RESEARCH PROJECTS

With the advances in process technology, the logic capacity and speed performance of FPGAs have increased significantly. This makes it possible to implement significant datapath functions and algorithms on the FPGA.

This has led to the development of platforms integrating processors and FPGAs, with the compute-intensive functions being off-loaded onto the FPGAs. Some of the research projects that have exploited FPGAs to achieve significant performance improvements are discussed in this section.

8.1 Programmable Active Memories [Bertin93]

The Programmable Active Memories (PAM) project is one of the first projects that proposed the use of an FPGA platform to accelerate software applications running on a host computer. The FPGA platform is used as a universal hardware co-processor closely coupled to a host computer. The host computer is used to download the configuration bit-stream into PAM. Conceptually in the architecture, the processor can access PAM like a RAM using read/write commands. Between the read and write instructions, the data is processed, hence the name Programmable Active Memories.

The DECPeRLe-1 is a specific implementation of the PAM architecture. The computational core is a 4 x 4 matrix of XC3090 FPGAs. Each FPGA in the array has direct connections to each of its four Manhattan neighbors. There are also common busses running in the vertical and horizontal directions. The granularity of the programming block, a four-input LUT, favors bit-level manipulation of the data.

This architecture has been used to implement RSA cryptosystems, DNA matching algorithms, and multi-standard video compression. Reported data show significant speedup over traditional processors.

8.2 Splash [Arnold93][Hoang93]

Splash is an attached parallel processor in which the computing elements are user-programmable FPGAs. The compute elements are arranged as a linear array. The Splash 2 implementation consists of a host computer, an interface board, and from one to sixteen Splash array boards. Each array board contains sixteen processing elements arranged in a linear array and fully connected using a crossbar switch. The processing elements are made of a XC4000 series FPGA and memory.

Single-instruction-multiple-data stream (SIMD) and systolic programming models are supported by Splash 2. The architecture was designed to accelerate computations that exhibit temporal and data parallelism. The

XC4000 series FPGA used in the processing element is programmable at the level of logic gates and flip-flops. Hence, this architecture is good for problems that require bit-level manipulation and lends well to systolic implementations.

The Splash 2 architecture has been used to search genetic databases. The problem of computing the edit distance between two genetic sequences was mapped as a systolic computation. The Splash 2 architecture was shown to outperform traditional computers by several orders of magnitude.

8.3 PRISM [Wazlowki93]

PRISM, which stands for Processor Reconfiguration through Instruction Set Metamorphosis, is a computer architecture consisting of a general-purpose processor and a reconfigurable platform. The PRISM-II implementation uses an AMD Am29050 processor coupled with an FPGA board made of three XC4010 devices. Compute-intensive functions are realized in the reconfigurable platform to obtain performance acceleration.

The compiler was built so that new operations can be synthesized automatically to match the computational characteristics of the application. The synthesized operations are then implemented on the reconfigurable platform to augment the functionality of the core-processor.

The automatic synthesis approach makes it attractive for use in a wide variety of applications. This approach was shown to yield significant speed-up over just using the core-processor. The choice of the bit-level reconfigurable platform helps it to perform better at applications requiring fine-grain manipulations.

8.4 ANT-on-YARDS [Tsutsui98]

This architecture combines tightly coupled FPGAs and a microprocessor. The implementation uses a 32-bit RISC processor (Hyperstone E-1) and an FPGA card made of XC4010 and MAX-9000 devices. These boards are coupled using an interconnection card that is programmable.

This architecture allows different styles of connections between the reconfigurable part and the processor: a bus, a direct interrupt, and a two-port SRAM channel. The bus style allows tight coupling between the two parts, with the reconfigurable board acting like a coprocessor. The interrupt driven mechanism enables the FPGA to interrupt and control the behavior of the processor. The third connection mechanism is aimed at scenarios where data communication between the processor and the FPGA occurs asynchronously and frequently. The first two schemes would result in local bus congestions and performance degradation in such a scenario. The two-port SRAM enables asynchronous data transfer between the two parts without one part influencing the performance of the other.

This architecture was used for implementing telecommunication data processing operations that require both high throughput and complex algorithms.

8.5 RaPiD [Ebeling96]

RaPiD is a linear array of coarse-grained functional units configured to form a mostly linear computational pipeline. The array of functional units is made of identical cells replicated to form a complete array. In the RaPiD-1 implementation, each of the cells is made of an integer multiplier, three integer ALUs, six general-purpose datapath and three small local memories. The interconnect is made of 16-bit segmented busses. The computational model proposed is a tightly coupled co-processor to a general-purpose processor with its own special path to memory.

The RaPiD architecture is aimed at regular, computation-intensive tasks like those found in digital signal processing. The logic block in RaPiD is coarser than those employed in the architectures discussed in the previous sections. Another interesting feature is the dynamic control available in RaPiD. In the traditional systolic array processing, the datapath is statically configured into a deep pipeline, and the data is streamed through it. RaPiD allows dynamic control to schedule different operations of the computation onto the datapath over time.

The coarse-grain programmability of the architecture favors arithmetic computations at the byte level that can take advantage of the fast nearest-neighbor connections available between the functional units. Preliminary results show impressive gains on DSP computations like matrix multiply, discrete cosine transforms, and motion estimation.

8.6 MATRIX [Mirsky96]

MATRIX is composed of an array of 8-bit functional units overlaid with a configurable network. Each functional unit is made of an 8-bit ALU and multiply unit, a 258 byte memory, and reduction control. The interconnection is made of three levels: nearest neighbor connection, length four bypass connection, and global lines.

MATRIX can be configured to operate in VLIW, SIMD, and MSIMD fashion. This architecture allows construction of master control out of the same pool of resources as array logic, avoiding the need for fixed control logic on each chip. The array can also be broken into units operating on different instructions. Mixed or varying granularities are handled by composing functional units and deploying instruction control. This is made possible because the datapath size and the assignment of control resources is not fixed.

The 8-bit granularity of the architecture is aimed at arithmetic computations that can utilize the ALU and the multiplier. Preliminary estimation shows a peak performance of 10Gops/second.

9 FPGA AND ENERGY CONSUMPTION

Most of the projects described in the previous section use FPGAs to accelerate performance. A natural progression from these projects is the use of FPGAs as a principal component in mainstream products. A major hurdle is the high power consumption of existing FPGA architectures. The same features that make an FPGA desirable are also responsible for making it extremely expensive in terms of energy. For example, it will be shown in Chapter 2 that the power consumption of an XC4085 chip running at a system frequency of 50 MHz is approximately 6W. At present, the high power dissipation is a limiting factor in energy sensitive domains.

The portable domain, with its numerous data sources, compute intensive tasks, and rapidly changing standards, can definitely benefit from the unique characteristics of the FPGA. The portable domain also places severe constraints on the power consumption of the hardware. The use of available FPGA architectures is not possible in these domains, where the current FPGA architectures will easily exceed the power budget by a few orders of magnitude.

To meet the demand for larger logic capacities, the size of FPGAs has been growing steadily over the past decade and is predicted to stay on this path. This has been made possible by staying at the forefront in terms of the process technology. The combined effect of smaller feature sizes and larger die area is the quadrupling of transistor count on a die. The resulting increase in power density and total power dissipation will have an adverse effect even in the power insensitive domains, because of the advanced packaging and the cooling techniques required.

10 CONCLUSION

Field-Programmable Gate Arrays have evolved considerably from their initial usage as just glue logic. The spatial realization of applications and the programmability of the architecture allow FPGAs to approach the efficiency of ASICs with the silicon reusability of general-purpose processors.

FPGAs are available with different methods of programming, interconnect architecture, and logic block functionality. Field-Programmable Gate Arrays are becoming popular as embedded components in computing platforms. The specific features will depend on the computing environment in which the FPGA is going to be deployed.

Chapter 2

POWER DISSIPATION IN FPGAS

1 INTRODUCTION

Field-Programmable Gate Arrays traditionally have been used in environments where their energy consumption was not critical. Present day portable devices have become more complex, and can take advantage of the programmability offered by the FPGA. This environment places stress on the energy efficiency of FPGAs, which is lacking in existing commercial architectures. Another factor that has gained importance is the power density of integrated circuits. With the reduction in feature size, the transistor density and transistor count per die have increased. This has resulted in an increase of power density, and the overall power dissipation per chip. This trend will continue, and has implications on the economics and technology of packaging these devices.

This chapter looks at the effect of process advances on the power dissipation of FPGAs. In the latter part of the chapter, the commercial FPGA architecture will be evaluated to isolate the effects of the different architectural components on the total energy consumption. This will help in identifying blocks that must be redesigned to improve the energy efficiency of the FPGA architecture.

2 TECHNOLOGY AND POWER

Improved speed performance, higher functional density, and the reduced cost per function afforded by deep-submicron processes are compelling reasons to design FPGAs using the most advanced process technology available. The minimum feature size has been steadily decreasing over the past decade.

The International Technology Roadmap for Semiconductors (ITRS) provides guidance for the semiconductor industry. This document is a collaborative effort of Semiconductor Industry Association, trade organizations of Europe (EECA), Korea (KSIA), Japan (EIAJ), and Taiwan

(TSIA). The ITRS provides a 15-year outlook on the major trends in the semiconductor industry. The most recent document, produced in 1999, predicts a minimum feature size of 35nm by the year 2014 [ITRS99].

The semiconductor outlook for the gate length and supply voltage is reproduced in Table 1 and Table 2. The predictions are broken into two groups: near terms years spanning 1998 to 2005 and long term years spanning 2008 to 2014. These data are based on ITRS1999, but there is an indication that the technology nodes will be pulled in by a year.

Table 1. Process Generations – Near Term Years [ITRS99]

Year	1998	1999	2000	2001	2002	2003	2004	2005
Gate Length (nm)	250	180	165	150	130	120	110	100
Supply Voltage (V)	2.5	1.8	1.8	1.5	1.5	1.5	1.2	1.2

Table 2. Process Generations – Long Term Years [ITRS99]

Year	2008	2011	2014
Gate Length (nm)	70	50	35
Supply Voltage (V)	0.9	0.6	0.6

Based on the technology roadmap from the ITRS [ITRS99] and FPGA vendors [Xilinx1] [Altera1], it is possible to estimate the logic capacity, speed performance, and power dissipation of the commercial FPGA architecture in the different process generations.

The reduction in feature size increases the logic density, making it possible to pack more computational power per unit area of silicon. The die area of future chips is predicted to remain at least constant, which means that the logic capacity of FPGAs will continue to increase. The improvement in process technology is accompanied by a reduction in circuit delay, and hence an increase in system clock frequency. Fig. 1 shows the trend in logic capacity and system clock frequencies of commercial FPGAs as a function of process technology to the year 2014. To estimate this trend, data from existing FPGA architectures and the roadmap from ITRS for ASIC generations are used.

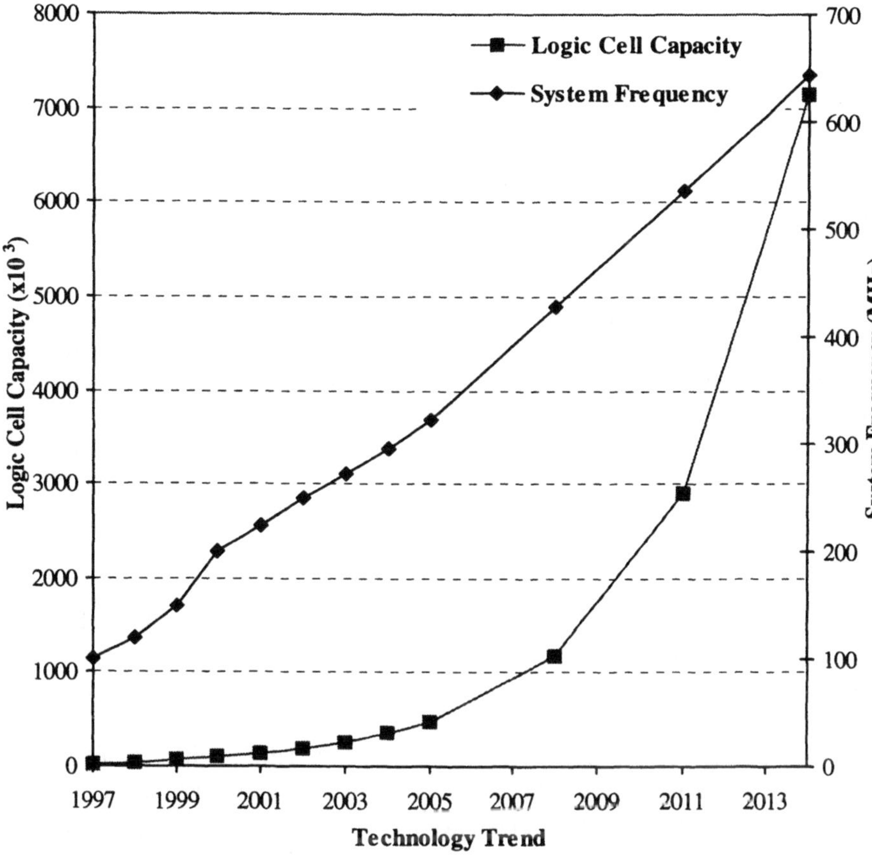

Figure 1. Logic Capacity and System Frequency of FPGAs

The estimation method can be explained using an example. Xilinx [Xilinx1] gives the logic capacity of its largest FPGA as 100,000 logic cells for the year 2000. Each logic cell is equivalent to a 4-input lookup table with an associated flip-flop. The ITRS roadmap [ITRS99] predicts an increase in transistor density for ASIC from 28Mtransistor/cm^2 in the year 2000 to 133Mtransistors/cm^2 in the year 2005. The size of the chip is predicted to remain constant over the same time period. Using these data, the logic capacity of the FPGAs can be estimated to increase to 475,000 logic cells by the year 2005.

$$LogicCapacity_{2005} = LogicCapacity_{2000} \cdot \frac{TransistorDensity_{2005}}{TransistorDensity_{2000}} \cdot \frac{DieSize_{2005}}{DieSize_{2000}}$$

$$LogicCapacity_{2005} = 100000 \cdot \frac{133}{28} \cdot 1 = 475000$$

A similar approach is used to estimate the system frequency. As the minimum feature size scales from 350nm to 35nm, the logic capacity of FPGAs is predicted to increase by more than two orders of magnitude, and the system frequency is expected to increase by a factor of six.

The gate oxide is scaled along with the feature size, which necessitates a corresponding scaling down of the supply voltage to prevent tunneling in the gate. The supply voltage is predicted to scale from 3.3V to 0.6V as the minimum feature size scales from 350nm to 35nm. The reduction in supply voltage, and hence signal swing, means that the energy per transition will reduce. System designers have been depending on this voltage scaling, and the reduced parasitics accompanying advanced processes, to obtain low power operation.

For a given function, the power dissipation will reduce with feature size. But, as shown in Fig. 1, the logic capacity per chip will increase. The higher integration, combined with the higher clock speed, will actually result in higher overall chip power. Fig. 2 tracks the overall power dissipation of an FPGA with process technology.

The power is estimated by assuming an array filled with 16-bit counters with an 85% utilization of the array. This is a benchmark used by FPGA vendors to estimate power dissipation [Altera97][Xilinx97]. The power dissipated in an XC4085XL chip running at 100MHz is computed to be 12.5W. This FPGA is implemented in a 350nm process with a supply voltage of 3.3V. Using this as a starting point, the power consumption of future generations is estimated based on the scaling of the voltage, logic capacity, system frequency and capacitive parasitics. The scaling of the voltage is obtained from Table 1 and Table 2. The change in the logic capacity and frequency are obtained from Fig. 1. The capacitive parasitics are assumed to scale linearly with the minimum feature size.

The change in power consumption from a 350nm technology to a 100nm technology is computed below as an example.

$$Power_{2005} = Power_{1997} \cdot \frac{Frequency_{2005}}{Frequency_{1997}} \cdot \frac{LogicCapacity_{2005}}{LogicCapacity_{1997}} \cdot \frac{Cap_{2005}}{Cap_{1997}} \cdot \frac{SupplyVoltage_{2005}^2}{SupplyVoltage_{1997}^2}$$

$$Power_{2005} = 12.5W \cdot \frac{322}{100} \cdot \frac{475{,}000}{10{,}000} \cdot \frac{0.1}{0.35} \cdot \left(\frac{1.2}{3.3}\right)^2 = 72.2W$$

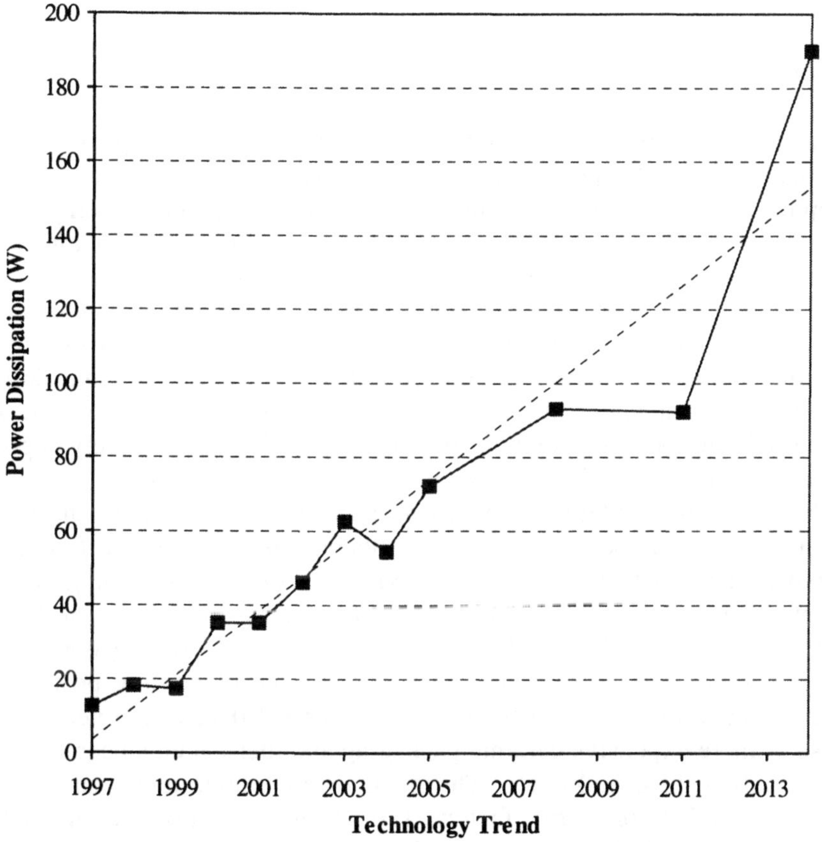

Figure 2. Power Dissipation of FPGAs

In the short term, the power dissipation is seen to increase at least linearly, despite scaling of the feature size and supply voltage.

3 IMPACT OF POWER DISSIPATION

As seen in the previous section, the power dissipation of the FPGA architecture will not be solved by the scaling of technology. The main impact of the increased power dissipation is in packaging and portability.

3.1 Packaging

There is a well-defined relationship between the junction temperature in an integrated circuit, the performance, and the reliability of the circuit. For example, the delay increases by 14% when the junction temperature increases from 85°C to 125°C [Xilinx97]. The reliability of an integrated circuit degrades as an exponential function of the junction temperature.

The main purpose of packaging and thermal control is to maintain the junction temperature within the functional and maximum allowable limits while maintaining acceptable circuit reliability. The functional limit defines the temperature at which the performance specifications can be met. The maximum limit is the temperature to which the device can be safely exposed without causing irreversible changes in operating characteristics.

To highlight the limitation imposed by packages, Fig. 3 shows the power dissipation of a commercial FPGA as a function of system clock frequency. The power dissipation is of an XC4085XL [Xilinx2] FPGA running at 3.3V, computed with an array utilization of 85%, and signal activity of 12.5%. Power handling capabilities of the readily available packaging options are also charted. As can be seen, the packages run out of cooling power well before the achievable system frequency of over 80MHz, which necessitates the use of more expensive ceramic packaging and thermal enhancements like forced cooling, heat sinks, etc. The main repercussions are economics of packaging, and poor form factor. The latter is important in mobile applications like cellular telephones, where the appliance has to be small, and worn on the body.

If the power dissipation of the chip follows the trend shown in Fig. 2, in the near future the power dissipation will be beyond the capability of existing packaging solutions. For example by the year 2005, the power dissipation will be ~70W, well beyond simple cooling techniques.

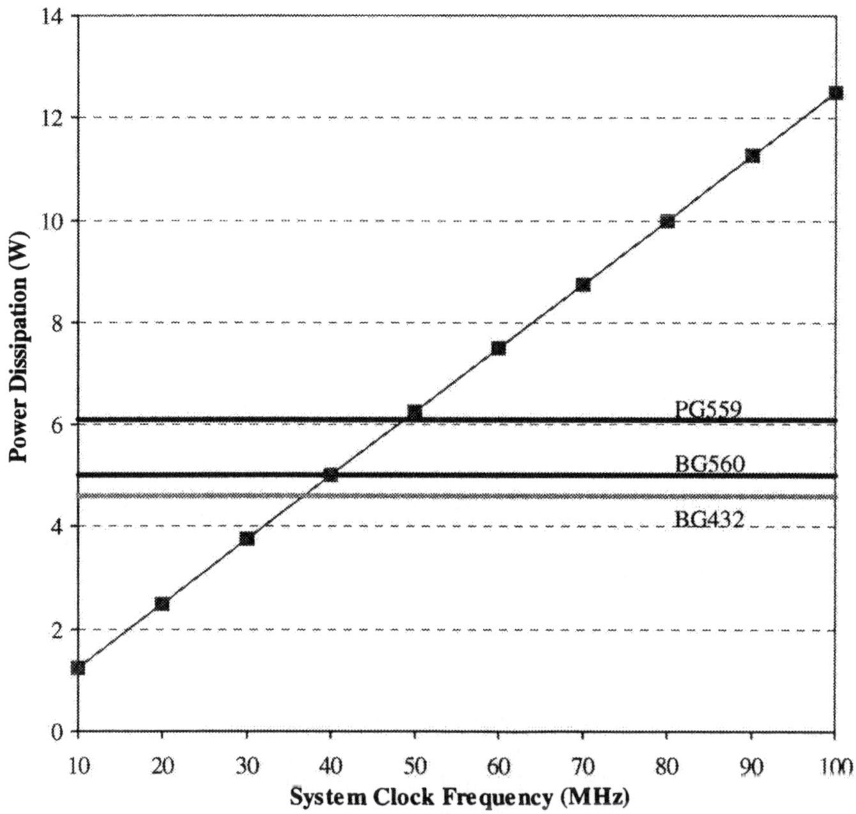

Figure 3. Power Dissipation as a Function of Speed for XC4085XL

3.2 Energy Efficiency and Portability

The poor form factor resulting from advanced thermal control is a handicap for portable applications. A more serious limitation is the energy efficiency of FPGAs.

One of the developing fields in which reconfigurable devices like FPGAs can be used is the cellular telephone market, or any portable environment which uses different data sources such as audio, video, etc. Communication devices, which must adhere to the different communication standards around the world, can also take advantage of reconfigurable architectures. These portable environments have stringent power constraints. In the case of cellular telephones, the power budget is of the order of tens of milliwatts, to maximize battery life.

As seen in Fig. 3, the FPGA dissipates power on the order of watts when running at tens of megahertz. Energy delivery devices have hit a plateau in their energy capacity, and a major breakthrough in this field is required if any more energy is to be extracted from them. In the meantime, aggressive energy saving techniques are needed in circuit design to make FPGAs a viable component in the portable environment

4 COMPONENTS OF POWER

A dramatic improvement in energy efficiency of FPGAs is required. An understanding of the energy breakdown in an FPGA is required to enable an efficient redesign process. Fig. 4 gives the energy breakdown of an XC4003 FPGA over a set of benchmark netlists [Kusse97].

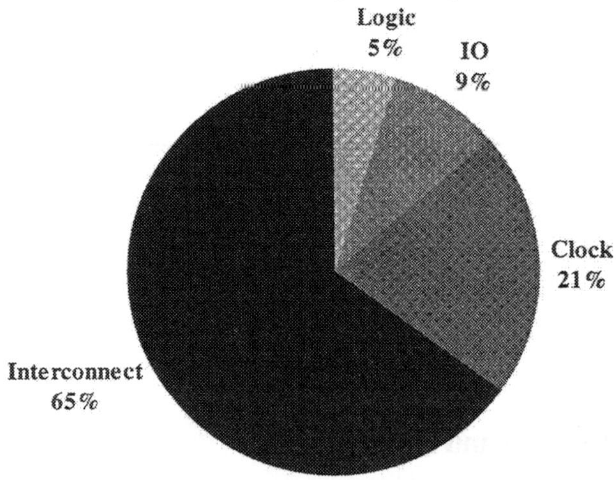

Figure 4. Power Breakdown in an XC4003 FPGA [Kusse97]

The majority of the power is dissipated in the interconnect. The next major component is the clock network. The logic block consumes only 5% of the total energy. This breakdown is not specific to the Xilinx FPGA, but is representative of most of the commercial FPGA architectures.

4.1 Interconnect Energy

The term "interconnect" includes all of the resources required to realize a connection between two logic blocks. As an example, consider the connection in a Symmetric Mesh architecture, illustrated in Fig. 5. The physical realization of the connection involves metal traces and programmable switches that have to be activated.

The interconnect architecture of the XC4000XL FPGA is similar to the Symmetric Mesh architecture. To obtain more information about the interconnect energy, a detailed power measurement of an XC4005XL FPGA was done. The XC4005XL FPGA is implemented in a 0.35μm process with a power supply of 3.3V.

It is possible to measure the energy dissipated in each of the single segment lines by selectively activating the switches in the connection box and the switch box. Table 3 gives the measured values.

Table 3. Measured Interconnect Energy of XC4005XL

Power Supply	3.3V
One CLB driving 9 single segments (Energy per transition cycle)	320pJ
One CLB with no load (Energy per transition cycle)	39pJ
Capacitance per segment	2.9pF

The energy of a single segment line is 39pJ. Assuming a full voltage swing on the line, this translates to a capacitance of 2.9pF. For this technology, the load of 2.9pF is considerable, equivalent to a fan-out of approximately 300 logic gates in a standard cell design. Since almost all the connections from an output pin have to go through at least one single segment line, the total switched capacitance is extremely high. This gives an idea as to why the interconnect structure dominates the total energy dissipation.

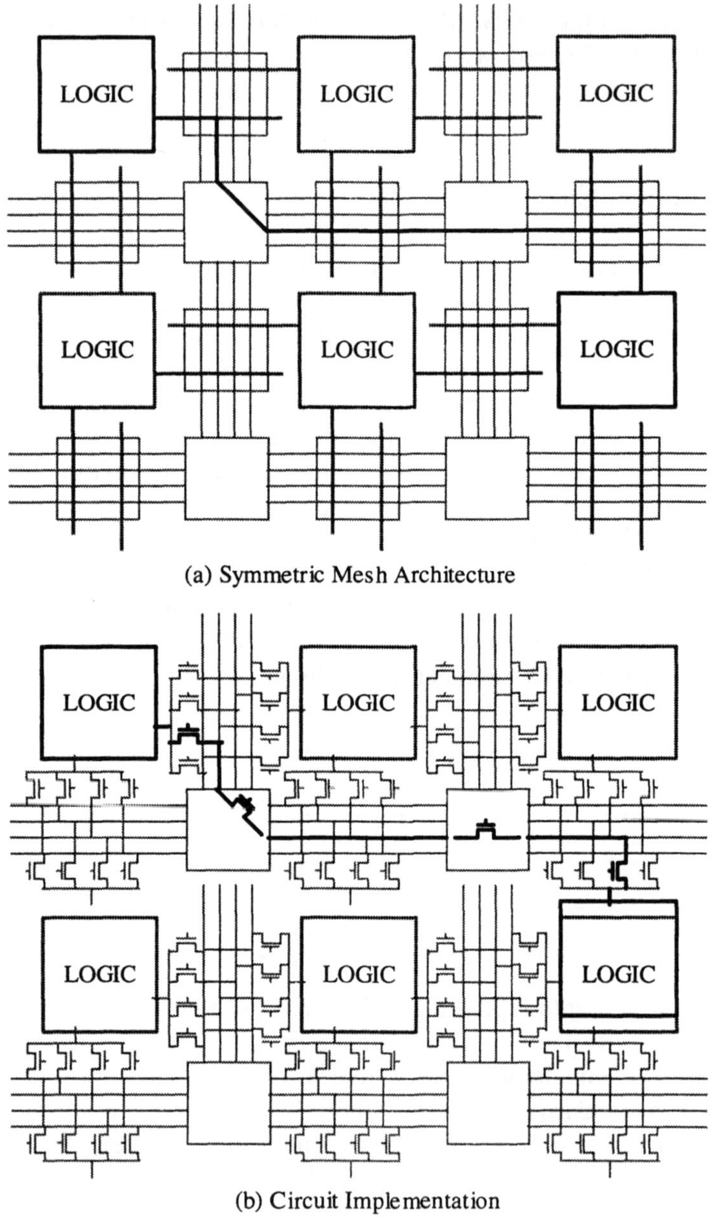

Figure 5. Routing in a Symmetric Mesh Architecture

Fig. 6 shows detailed information about the connections possible to a track of the single segment length interconnect in XC4000. The connections are from the input/output pins of the logic blocks, the switch boxes, and the long lines.

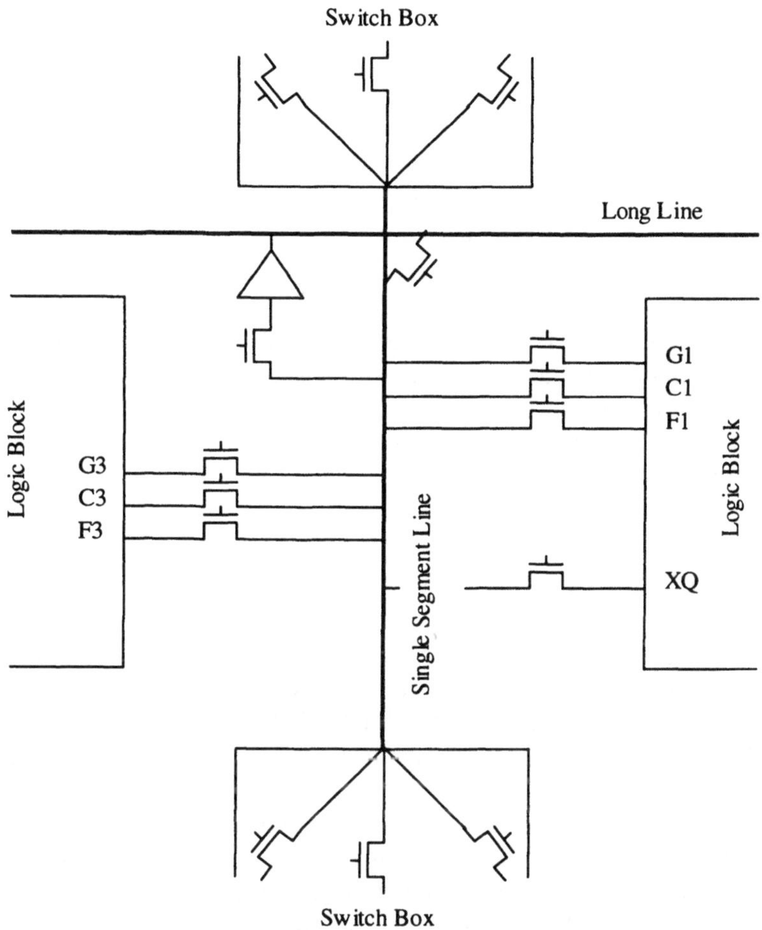

Figure 6. Single Segment Line in XC4000 Architecture

The capacitance on the line is from the metal track spanning one logic block, and from the diffusion capacitances of the pass transistors connected to this metal track. Using reported data on the logic block size [Wittig96], and with scaling for technology, the length of a single segment line is ~300μm in a 0.35μm technology process. Using metal capacitance data, the contribution from the metal line is ~40fF. The rest of the load is from the diffusion capacitance of the NMOS transistors connected to the single line.

This can be reduced by either decreasing the number of switches accessing the line, or by making the transistors smaller. The number of switches can be decreased by reducing the flexibility of the switch box and the connection box, and by reducing the width of the routing channel. Doing so adversely affects the routing efficiency of the interconnect architecture. Hence, any modification of the flexibility has to be accompanied by an evaluation of the routing efficiency of the entire architecture.

The interconnect path in an FPGA can be modeled as an RC chain. The resistance of the series transistors contributes to the R, while the diffusion capacitance of the NMOS transistors in the path contributes to the C. By reducing the width of the switch, the R of the series path increases, and the speed performance suffers. The commercial FPGA architecture typically uses wide transistors to reduce the series resistance of the switches. If the architecture can be modified so that the number of series transistors can be reduced, then the width of the switches can be reduced while maintaining the speed performance. The relationship is explored in Chapter 4.

5 CLOCK ENERGY

The next major contributor to the total energy is the clock. To get a better idea of the clock energy, a simple H-tree clock distribution tree is designed for a 16 x 16 array of logic blocks. Each of the logic blocks has two flip-flops. The clock distribution is shown in Fig. 7. Each of the clock regions is of size 4 x 4.

Typically in all FPGAs, flip-flops are provided in each logic block to register the output. It has been shown that the availability of flip-flops in each logic block improves the utilization of the array, and leads to a better area efficiency [Rose90a]. A side effect of this architectural decision is that the clock has to be distributed over the entire array to supply the sparse distribution of flip-flops. This results in a relatively large cost for the clock distribution network.

Table 4 gives the energy breakdown of the total clock distribution. The distribution tree takes into account the capacitance of the H-tree, as well as the overhead of the distributed buffers. The tile energy represents the clock energy consumed in the leaf cells. This includes the clock distribution in the tile, as well as the local clock buffer.

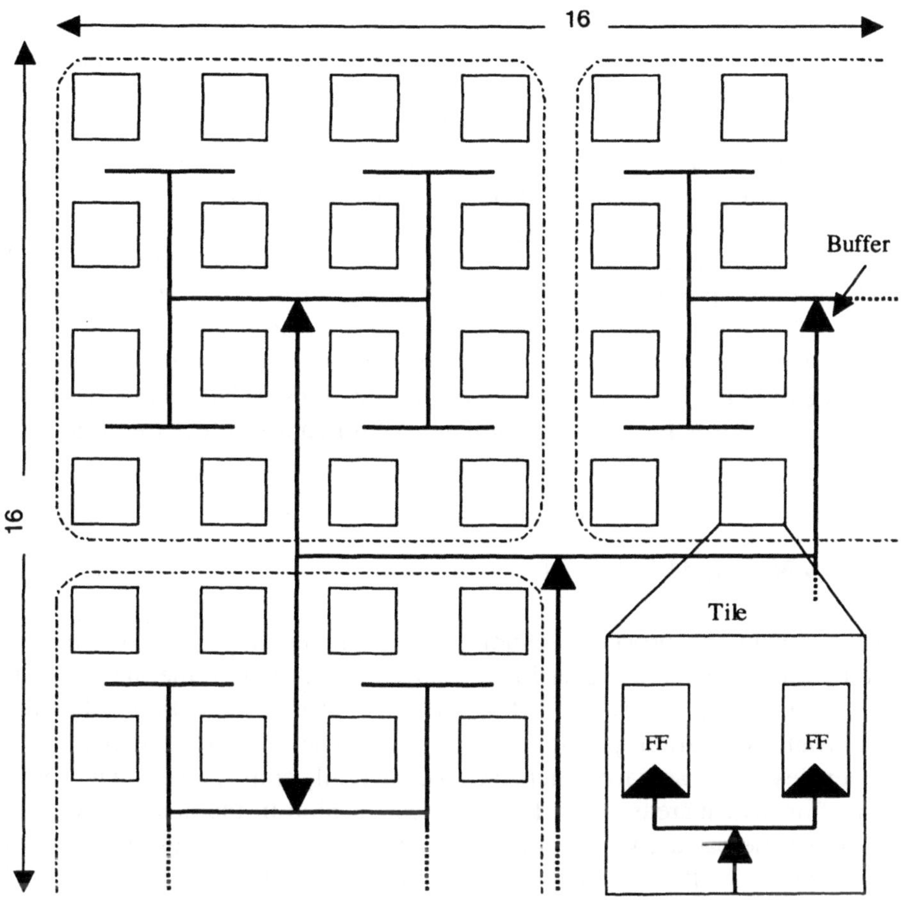

Figure 7. Clock Distribution Tree for a 16 x 16 Array

Table 4. Energy Components of Clock

	Global Distribution	Tile
All the tiles are active	40%	60%
30% of the flip-flops are active	70%	30%

When all of the tiles are active, the entire global distribution network, as well as the local distribution, is active. In such a situation, the global distribution accounts for 40% of the total clock energy. For typical applications, not all of the tiles' outputs need to be registered. In such a situation, when the clocked tiles are scattered sparsely over the array, the entire global distribution is still active while only a fraction of the local distribution networks are active. If a usage of 30% is assumed, the global distribution now accounts for 70% of the total clock energy. This indicates that reduction of the global energy will be rewarded with a significant reduction in the total clock energy.

Another interesting factor is that the contribution of the flip-flops to the clock energy is minimal. Consider the example of the 16 x 16 array. In this array, 512 flip-flops are distributed over an area of 4.8mm x 4.8mm. The clock network that has to be distributed over this entire area will certainly dominate over the input capacitance of the flip-flops. What this means is that the complexity of the flip-flops will not significantly affect the total clock energy.

6 CONCLUSION

Advances in the process technology will aid in the quest for improved speed performance and higher logic capacity. Even though the move into the deep-submicron processes and the associated voltage reduction will reduce the power dissipation per logic gate, thc higher transistor densities and frequencies will actually increase the overall chip power. The power dissipation will increase in at least a linear fashion in the next decade. This will affect the packaging requirements, and have adverse impact on the cost performance. A dramatic improvement in the energy efficiency has to be achieved to make FPGAs viable in the portable computing domain.

Detailed energy analysis based on existing FPGA architectures shows that the dominant components which have to be redesigned are the interconnect architecture and the clock distribution network.

The dominant component in the interconnect is the diffusion capacitance of switches used to provide the connectivity in the array. Since the number of switches is a direct result of the design of the interconnect architecture, the architecture has to be redesigned from an energy perspective. Since the interconnect architecture has a direct impact on the routing resources available for routing the nets, the redesigning effort has to be done without adversely affecting the routability of the FPGA architecture.

It is shown that for the clock energy, the dominant component is actually the distribution network, and not the load presented by the flip-flops. Hence, the distribution network has to be targeted first to reduce the clock energy.

Chapter 3

EXPLORATION ENVIRONMENT

1 INTRODUCTION

The logic and routing resources in an FPGA are limited. The logic capacity is determined by the number of logic blocks and the interconnect capacity is determined by the interconnect architecture. It is not possible to implement a function on a given array if the gate count required for the function exceeds the logic capacity of the array, or if the required connections cannot be completely supported by the available routing resources.

The logic capacity, quantified in terms of the gate count or k-input LUTs, can be increased by increasing the number of logic blocks. Increasing the routing capacity is more involved. The capacity of the interconnect architecture is affected by the number of tracks in the routing channel, the flexibility of the connection box, and the flexibility of the switch box. Changing the flexibility of the connection box and the switch box, and modifying the number of tracks has a direct impact on the number of routing switches required. Increasing the number of routing switches per track degrades the energy and speed performance of the interconnect due to the increase in the parasitic capacitance from the switches. Since the switches require configuration memory for programming, the increase in the number of memory cells will increase the total area. Hence, the architecture optimization is a tradeoff between architecture flexibility, area, speed, and energy performance.

This chapter describes the flow that is used to implement benchmark applications onto the target architectures. The exploration environment aids in placing and routing a netlist of LUTs onto different architectures. This makes it possible to obtain performance costs of different architectures.

2 RELATED RESEARCH

Considerable work has been done in the academia to develop placement and routing tools to evaluate different FPGA architectures. The placement problem in the FPGA environment is similar to that in VLSI. Therefore, these techniques [Sechen87] have been adapted for use in FPGA. Most of the work has been done for developing the router.

The Coarse Graph Expansion (CGE) [Brown92] router has been developed mainly for use in Symmetric Mesh architectures. Most routers are inherently sequential, which means that when one net is being routed they are unable to consider the effects on other nets. CGE attempts to bypass this limitation by evaluating all the nets at the same time. This router accepts coarse graphs from a global router and expands it into detailed routes. In the first step, all the possible paths for each net are elaborated based on the coarse graph. During the next step, all the paths are compared, and specific paths for each net is chosen based on a cost function.

SEGA [Lemieux93] improves on CGE by not only focusing on achieving full routability, but also addresses the allocation of wire segments to connections to match the length of the segment to the length of the connection. This tool takes into consideration the fact that the routing switches result in performance degradation, and attempts to use longer wires with fewer routing switches. SEGA supports segmented symmetrical architectures.

The Versatile Placement and Route (VPR) [Betz97b] encapsulates the entire placement and routing flow. VPR is targeted at Symmetric Mesh architectures. The tool allows the specification of the size of the logic block, pin accessibility, routing widths of the channel, and the distribution of the channel widths. Simulated annealing is used for the placement step. The router is based on the Pathfinder negotiated congestion algorithm [Ebeling95]. Reported data show superior performance compared to other FPGA placement and routing tools.

Alexander et al. propose an architecture-independent approach to FPGA routing based on multi-weighted graphs [Alexander94][Alexander96]. This allows simultaneous optimization of multiple objectives under a smooth designer controlled tradeoff. This method is based on a multi-weighted graph formulation of the FPGA architecture. This allows the tool to be completely independent of the specific architecture. The routing problem is posed as a minimum spanning tree problem.

3 EVALUATION FLOW

The evaluation flow is shown in Fig. 1. It encapsulates a complete flow from a logic netlist to the implementation on the target architecture.

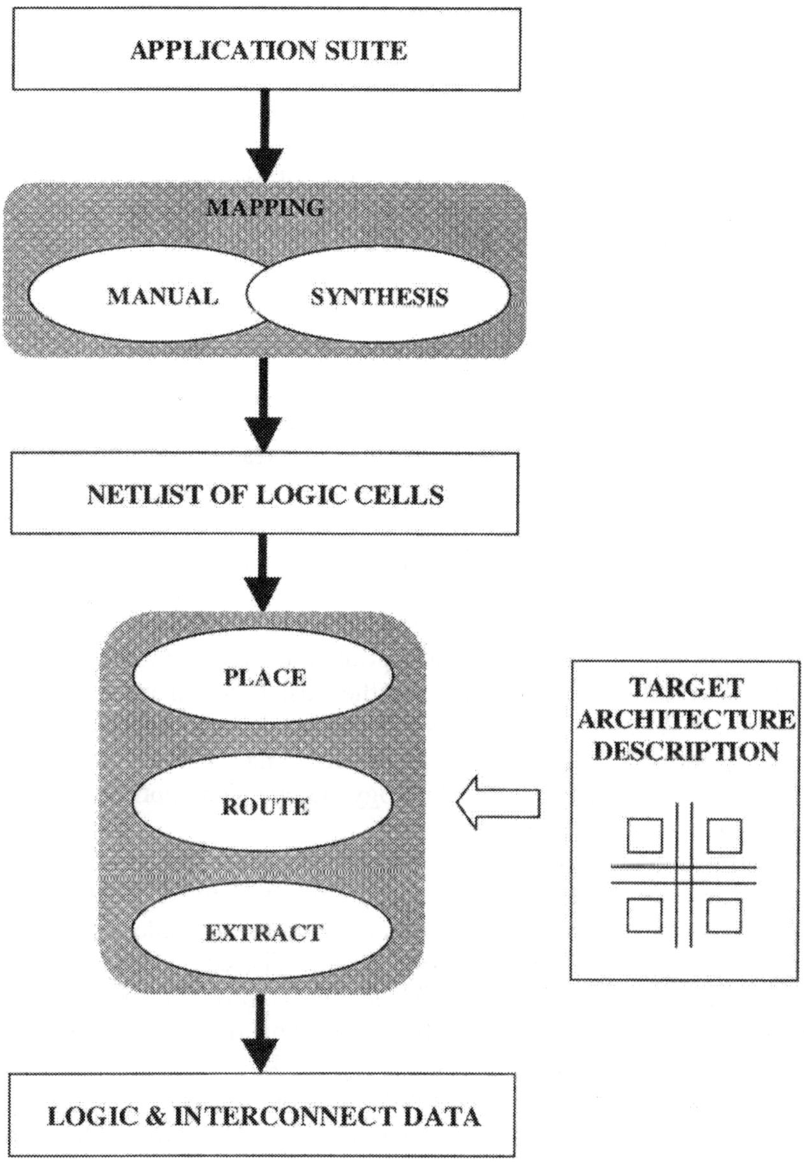

Figure 1. Evaluation Flow

The applications are mapped to a netlist of lookup tables (LUTs). The core of the tool is a placement and routing tool, which is used to implement the netlist in the target architecture. The target architecture description is treated as another input to the tool, and can be changed to evaluate architectural modifications. After successful implementation of the application, all necessary data can be extracted to evaluate the architecture from an energy and speed performance perspective.

The benchmark suite is a combination of data-path and random logic circuits. They are drawn from the MCNC benchmark suite and from functions encountered in the DSP domain.

4 MAPPING

Mapping is the process of translating the circuit description to a netlist of LUTs. Fig. 2 illustrates the steps involved in a traditional automated mapping flow.

The first step is the logic optimization phase, where the Boolean expressions are optimized. The purpose of this step is to reduce the complexity of the network based on a cost function. This is typically achieved by removing redundancies and common sub-expressions. The *don't care* conditions in the logic description are also exploited to reduce the complexity of the network. Usually the cost function used to evaluate the network is the number of literals for the local functions in each node. The resulting network is logically equivalent to the initial logical description. This step is referred to as "technology independent optimization", since it does not consider the final architecture.

The next step is technology mapping. At this step, the optimized network is mapped onto the target architecture. This step looks at the components available in the target architecture, and tries to implement logic in the optimized network using these components. The cost functions usually used are area and delay. Typical FPGA synthesis programs use a cell library, which is annotated with the cost functions. This does not take full advantage of the lookup tables provided in the logic blocks.

MIS-FPGA [Murgai90] and FlowMap [Cong94] can be used to map the optimized network to a netlist of lookup tables. These tools allow the size of the LUT to be varied. This is useful in exploring the effect of the logic block granularity. In this work, MIS-FPGA is used in the mapping step.

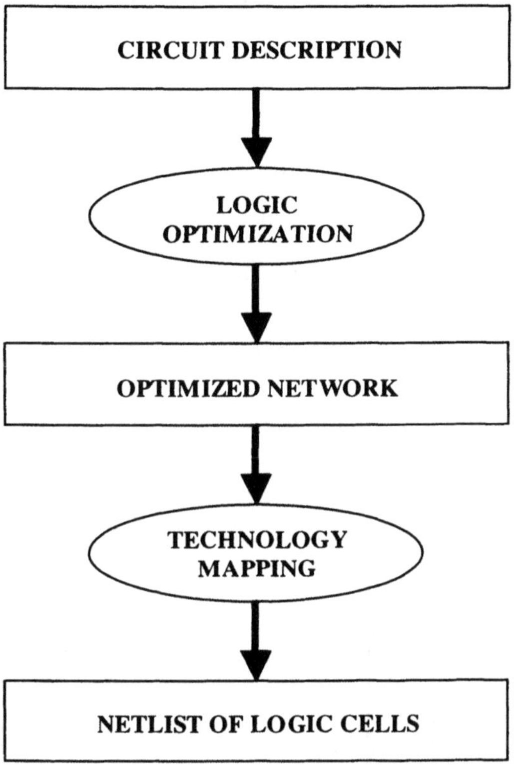

Figure 2. Logic Mapping

Manual mapping to a netlist of LUTs is done for critical functions and applications to take advantage of the specific characteristics of the logic block structure.

5 ARCHITECTURE REPRESENTATION

Almost all of the available placement and routing tools are closely tied to specific architectures with only limited access to architectural specifications. This makes it possible to speed-up the placement and routing step, but is not amenable to architectural exploration.

To redesign the FPGA architecture, the designer should have complete flexibility in specifying the characteristics of the target architecture. This is made possible by treating the target architecture description as an input to the evaluation flow, rather than hard coding the target information in the placement and routing tools. The main information the architecture description has to convey are:

- Amount of logic and routing resources available.
- Connectivity available in the routing architecture.
- Cost (energy and delay) of the routing resources.

All of the above information can be encapsulated by representing the FPGA architecture as a weighted graph. Fig. 3 shows how a simple logic block and its associated routing channel is represented in the form of a graph.

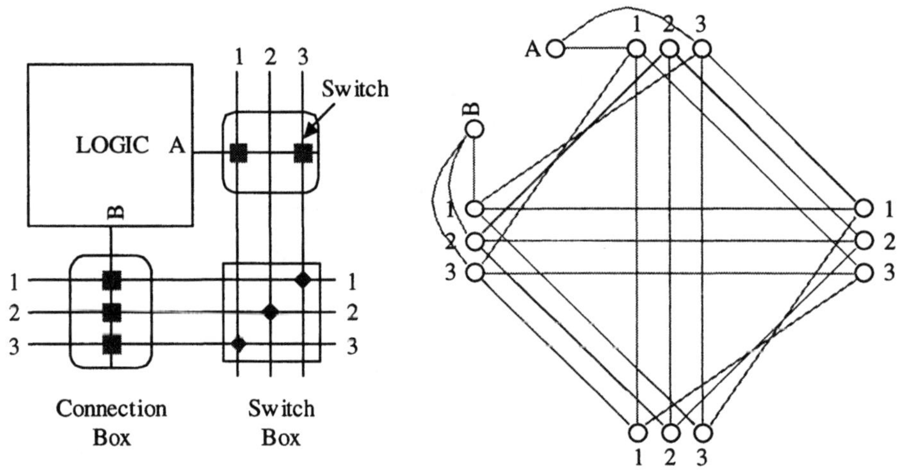

(a) Sub-block in a Symmetric Mesh Array

(b) Graph representation

Figure 3. Graph Representation of FPGA Architecture

Fig. 3(a) shows a portion of an array, with a logic block accessing the vertical and horizontal routing channels through pins A and B. The shaded squares represent switches in the connection box, and hence the connectivity available. Pin A can only access tracks 1 and 3 in the vertical routing channel, while pin B can access all the tracks in the horizontal routing channel. The switch box is similar to the Xilinx switch box, and each of the tracks can connect to the corresponding tracks on the other three sides.

```
////////////IOPAD(0,0)////////////
//LOGIC BLOCK
0.0.LOGIC.0.0.0 9C 0.0.WIRE.1.0.0 300C 0.0.WIRE.1.0.1 300C 0.0.WIRE.1.0.2 300C
0.0.WIRE.1.0.3 300C 0.0.WIRE.1.0.4 300C
0.0.LOGIC.0.0.1 9C 0.0.WIRE.1.0.0 300C 0.0.WIRE.1.0.1 300C 0.0.WIRE.1.0.2 300C
0.0.WIRE.1.0.3 300C 0.0.WIRE.1.0.4 300C
0.0.LOGIC.0.0.2 9C 0.0.WIRE.1.0.0 300C 0.0.WIRE.1.0.1 300C 0.0.WIRE.1.0.2 300C
0.0.WIRE.1.0.3 300C 0.0.WIRE.1.0.4 300C

//ROUTING CHANNEL
0.0.WIRE.1.0.0 10C 0.0.LOGIC.0.0.1 300C 0.0.LOGIC.0.0.2 300C 0.0.LOGIC.0.0.3
300C 1.0.EQVN.0.0.2 300C 1.0.EQVN.0.0.6 300C 0.1.WIRE.1.0.0 10C
1.1.WIRE.0.1.0 10C 1.0.WIRE.0.1.0 10C
0.0.WIRE.1.0.1 10C 0.0.LOGIC.0.0.1 300C 0.0.LOGIC.0.0.2 300C 0.0.LOGIC.0.0.3
300C 1.0.EQVN.0.0.2 300C 1.0.EQVN.0.0.6 300C 0.1.WIRE.1.0.1 10C .
.
.
////////////TILE(1,0)////////////
//LOGIC BLOCK
1.0.LOGIC.0.0.0 9C 1.0.EQVN.0.0.0 300C 1.0.EQVN.0.0.1 300C 1.0.EQVN.0.0.2
300C
1.0.LOGIC.0.0.1 9C 1.0.EQVN.0.0.0 300C 1.0.EQVN.0.0.1 300C 1.0.EQVN.0.0.2
300C
1.0.LOGIC.0.0.2 9C 1.0.EQVN.0.0.0 300C 1.0.EQVN.0.0.1 300C 1.0.EQVN.0.0.2
300C
.
.
//PINS
1.0.EQVN.0.0.0 0C 1.0.LOGIC.0.0.0 300C 1.0.LOGIC.0.0.1 300C 1.0.LOGIC.0.0.2
300C 2.0.EQVN.0.0.5 50C 1.1.EQVN.0.0.5 50C 2.1.EQVN.0.0.5 50C 1.1.WIRE.0.1.0
300C
.
.
.
//GROUPING
GROUP 0 1.0.WIRE.0.1.0 1.0.WIRE.0.1.1 1.0.WIRE.0.1.2 1.0.WIRE.0.1.3
1.0.WIRE.0.1.4
.
```

Figure 4. Sample Architecture Description

This structure is represented as a graph in Fig. 3(b). Nodes in the graph represent the pins and the routing tracks. Arcs connecting the corresponding nodes represent the possibility of a connection. For example, pin A can only connect to tracks 1 and 3 in the vertical routing channel. This is represented by arcs connecting the node representing the pin A to the nodes representing tracks 1 and 3; there is no arc between pin A and track 2. By using a weighted graph representation, it is possible to model the cost of the connecting resources with weights on the arcs. Fig. 4 is a sample architecture description file.

The weighted graph representation has the advantage that the routing problem can be tackled with existing graph algorithms. However, this method of decoupling the placement and routing tool from the architecture incurs the penalty of slower execution of these steps.

6 PLACEMENT

In a traditional VLSI design flow, placement is the process of placing blocks so as to minimize total chip area while ensuring that the critical nets can be routed within their timing budget. In an FPGA, the total area is fixed, and the placement step reduces to the problem of assigning the LUTs in the netlist to the available logic blocks in the array while working within the utilization and performance constraints. An important characteristic of the FPGA architecture is that the speed and energy performance are dominated by the interconnect, making the placement and routing steps extremely important.

One of the most common FPGA architectures is the Symmetric Mesh, and most of the tools developed [Betz97a] and architectural parameters studied, were based on this simple structure. For the Symmetric Mesh architecture, the cost of connecting pins is estimated using an approach similar to that used in VLSI system placement. The geometrical bounding box of the pins involved in the net is used as an indicator of the cost. This method of cost estimation becomes inaccurate when the interconnect architecture differs from the Symmetric Mesh structure. For example, for the Symmetric Mesh architecture in Fig. 5(a), the cost of connecting any two logic blocks is proportional to the geometric distance between the logic blocks. In Fig. 5(b), however, the regular single segments have been augmented with longer wires, which can provide low cost connections between logic blocks that are further apart. It was shown in the previous chapter that the cost of the connection is dominated by the resistance and the capacitance of the switches, rather than by the metal traces. In this situation, the earlier assumption of cost as a function of the geometrical distance no longer holds true. In commercial tools, this has been taken care of by hard-coding the architecture-dependent costs to the placement tool. This does not lend itself well to architecture modifications and experimentation. In this flow, the cost of connecting blocks is extracted from the weighted graph representation of the target architecture and is used in the placement algorithm.

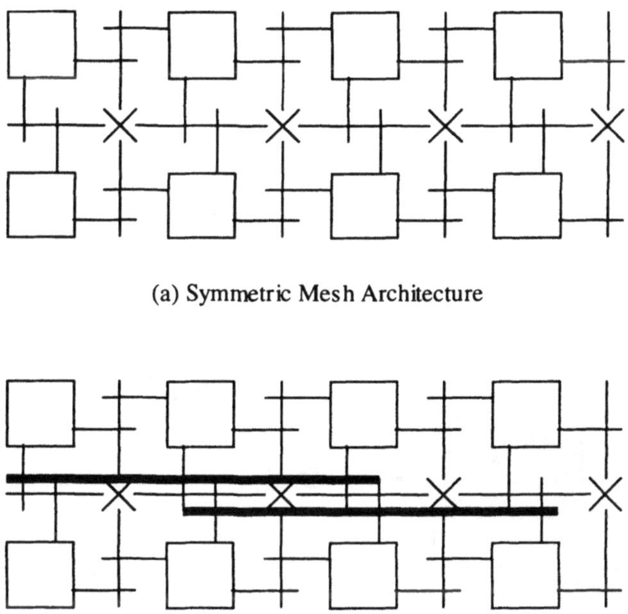

(a) Symmetric Mesh Architecture

(b) Modified Symmetric Mesh Architecture

Figure 5. FPGA Architectures

6.1 Placement Routine

Fig. 6 shows the different steps involved in the placement routine. The target architecture is described in the form of a weighted graph as described earlier. The circuit netlist is a connectivity graph, which describes the connections required between the different LUTs. The goal of the placement routine is to place the LUTs optimally from the netlist on the available logic blocks in the target architecture to minimize the interconnect cost. The optimization tool used is simulated annealing [Kirkpatrick83]. The cost of connections is described in the Distance Table.

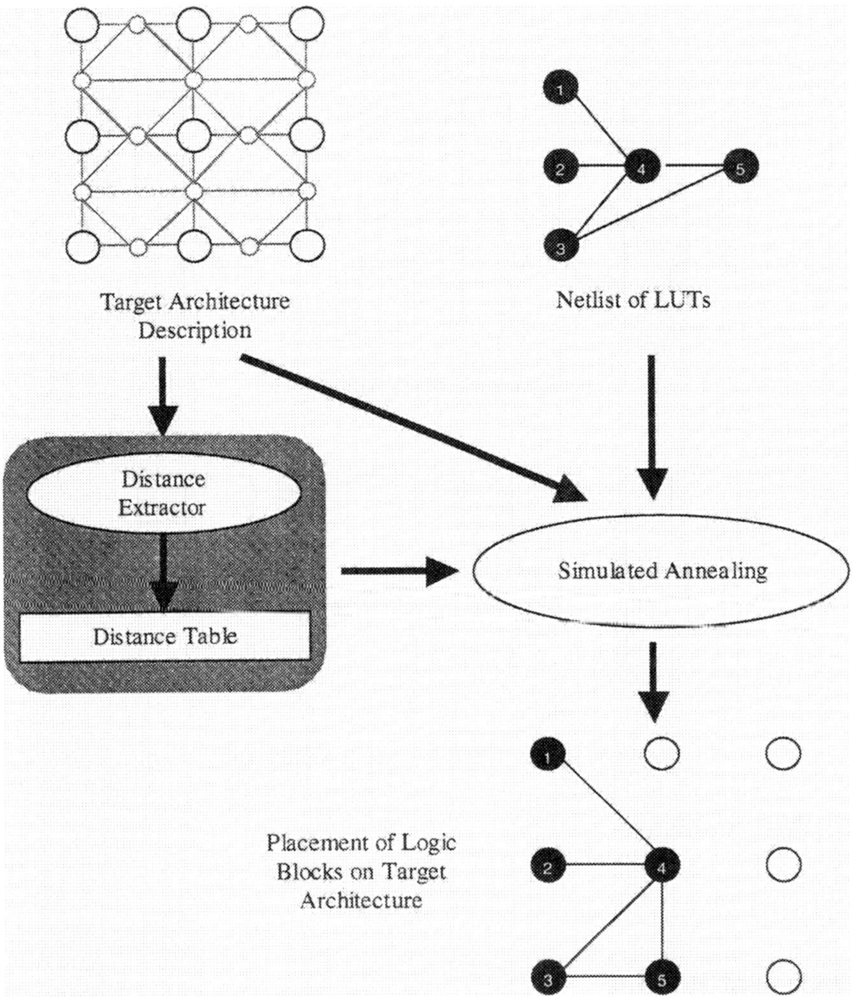

Figure 6. Placement Flow

6.1.1 Distance Table

The geometrical distance between logic blocks is a poor estimation of the cost of connecting the blocks. The connection cost is important since the optimization tool uses it to evaluate each placement option. The weighted graph architecture representation has the advantage that it encapsulates the cost of each of the routing resources.

Using the weighted graph representation, it is possible to provide the cost information required for the optimization tool. The costs can be obtained by routing a path between the logic blocks involved using any shortest path algorithm [Dijkstra59][Bellman58][Ford62][Cormen90]. The computational complexity of these algorithms is a function of the graph size. This, combined with the fact that the cost needs to be computed often at each temperature step of the simulated annealing algorithm, makes this approach computationally prohibitive. The implemented flow uses a lookup table approach. The distance between all pairs of logic blocks is computed once, and stored in the "Distance Table", as shown in Fig. 7.

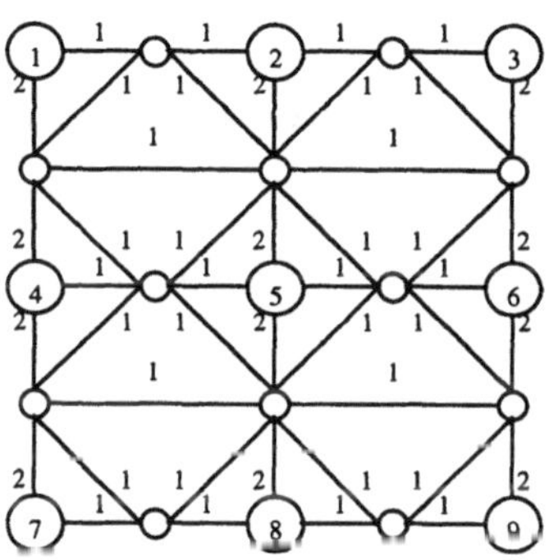

Weighted Graph

	1	2	3	4	5	6	7	8
9	6	6	6	4	4	4	4	2
8	6	6	6	4	4	4	2	
7	6	6	6	4	4	4		
6	4	4	4	4	2			
5	4	4	4	2				
4	4	4	4					
3	4	2						
2	2							

Distance Table

Figure 7. Distance Table of the Target Architecture

During simulated annealing, this table is used to compute the cost of the interconnect. Dijkstra's single source shortest path algorithm [Dijkstra59] is used for generating this distance template. The complexity of this algorithm is $O(|E|+|V|\lg|V|)$, where $|V|$ is the number of nodes, and $|E|$ is the number of edges in the graph, $G(V,E)$. The size of the distance table is proportional to the square of the size of the array. For example, the table for a 16 x 16 array has 496 entries.

6.1.2 Simulated Annealing

Simulated annealing is a versatile optimization technique widely used in standard cell placement packages, as well as in numerous other applications with highly nonlinear functions and non-convex or unconnected solution spaces. Specifically, simulated annealing is able to process cost functions with arbitrary degrees of nonlinearities and discontinuities, and to statistically guarantee a solution as long as the system "temperature" is decreased sufficiently slowly [Ingber93].

These properties make simulated annealing a very appealing optimization engine for the FPGA cell placement problem. The general structure of the simulated annealing algorithm applied to cell placement problems falls within the general category of "probabilistic hill-climbing algorithms," outlined below [Sechen85]:

```
T = T0;
X = generate(X0);     /* X0 is typically a randomly chosen
initial state */
while("stopping criterion" is not satisfied) {
   while("inner loop criterion" is not satisfied) {
       Xp = generate(X);
      if(accept(c(Xp), c(X), T)), X = Xp;
   }
T = update(T);
}
```

The functions **generate()**, **accept()** and **update()** represent the core features of the simulated annealing algorithm: the process of proposing a new circuit state based upon previous states and the current temperature of the system, the probabilistic acceptance test, and process by which the system temperature is decreased (annealed).

The initial placement is randomly chosen. Then the contents of two logic locations are swapped, and the change in cost is computed. All moves that reduce the total cost are accepted. Moves that result in an increase of cost are accepted with a probability that is dependent on the system temperature, T.

The acceptance probability is given by the Boltzman function, $e^{\frac{-\Delta C}{T}}$, where ΔC is the increase in cost.

The rate at which the temperature is decreased is critical to the success of the algorithm. Simulated annealing will converge to the global local minimum with probability 1.0 only if the temperature is decreased sufficiently slowly. In practice, almost all cell placement packages violate this constraint by not generating enough states according to the inner loop criterion at a given temperature T, or by simply decreasing the temperature too rapidly. Although the resulting simulated quenching does not guarantee convergence, the results achieved in practice are usually quite good. T is decreased according to $a(T)\ T$, where $a(T)$ is less than but very close to 1, typically 0.96 to 0.97.

As the system cools, the chances of accepting a move that results in a higher total cost reduce. Due to this, the percentage of swaps that are accepted reduces with temperature. Lam, et al. [Lam88] showed that the optimum acceptance rate of proposed new states is 44 percent. In the algorithm, the acceptance rate is controlled by controlling the window from which new swap locations are chosen. As the system cools, the size of the window grows smaller, to maintain this acceptance ratio. Swartz, et al. [Swartz90] improved on this algorithm so that it is possible to predict at the beginning of the run, when the algorithm will end. The theoretically derived annealing schedule uses an exponentially decreasing range window, while the temperature is controlled to maintain the acceptance ratio.

In this work, the algorithm proposed in [Lam88] is employed. The range window is defined in terms of the cost of the connections. In the **generate()** function, the first location is selected randomly, while the second location is randomly chosen, so that the cost is within the window. These data are available directly from the distance table.

7 ROUTING

Routing is the process of realizing the connections between the logic blocks on the target architecture after the position of the logic blocks have been fixed. Routing on FPGAs differs from the VLSI routing problem in the sense that the resources are fixed. Since the number of tracks in a channel is fixed and the possible connections between tracks are fixed, routing is a critical step in FPGAs.

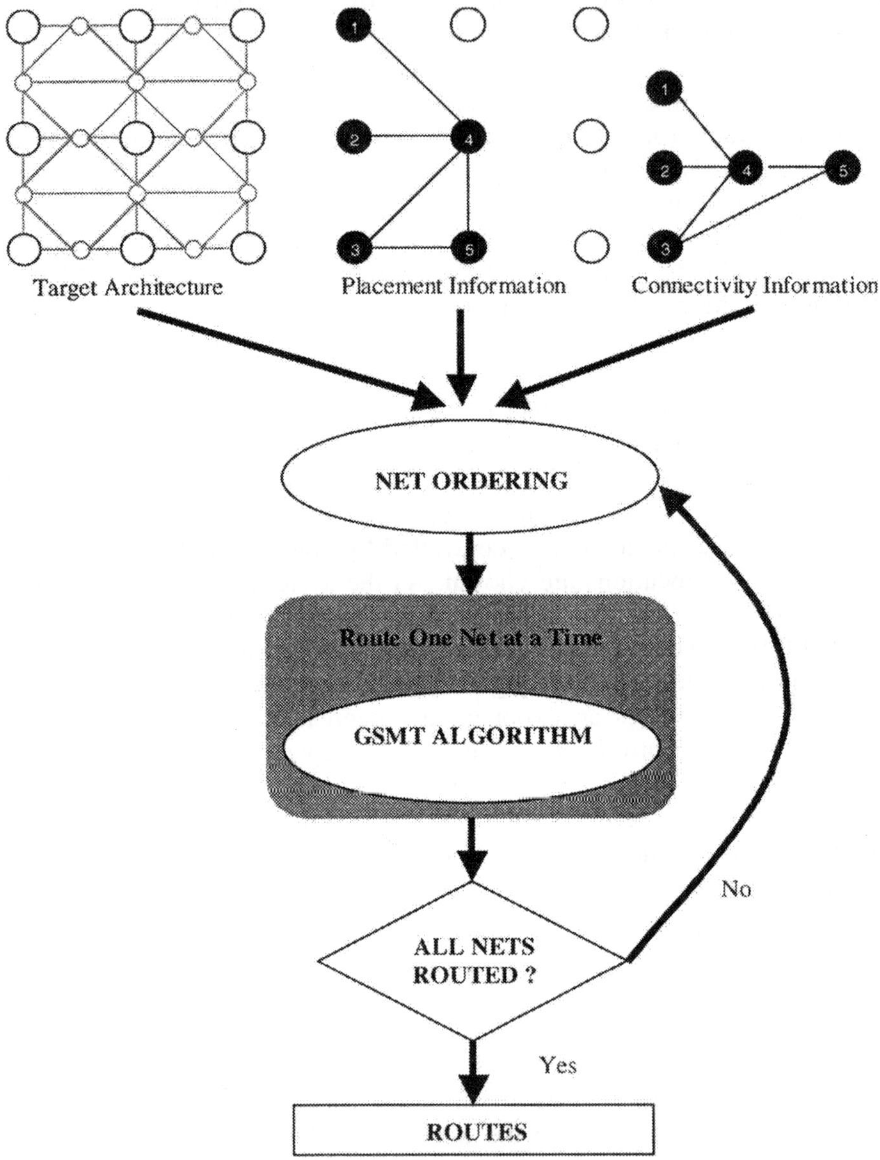

Figure 8. Routing Flow

One of the advantages of the graph-based approach is the availability of a number of routing algorithms that can operate on weighted graphs. In this work, the routing is treated as a Graph Steiner Minimal Tree (GSMT) problem. As long as the architecture is described as a weighted graph, the relevant graph algorithms can be employed irrespective of the irregularity or non-uniformity of the routing resources. The GSMT is a graph version of the Steiner Minimal Tree (SMT) problem. It is illustrated in Fig. 9.

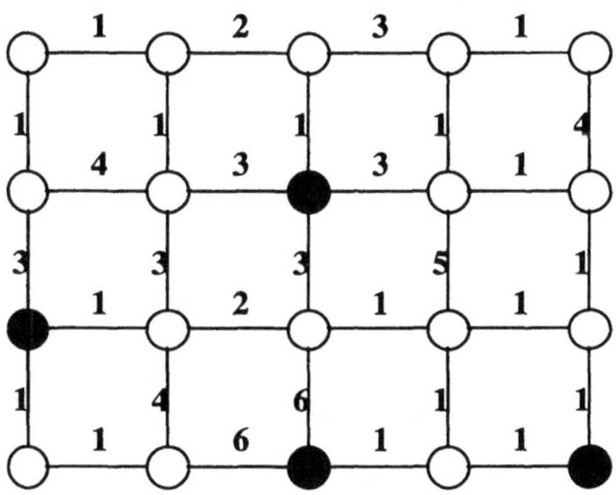

(a) Nodes to be Connected in a Weighted Graph

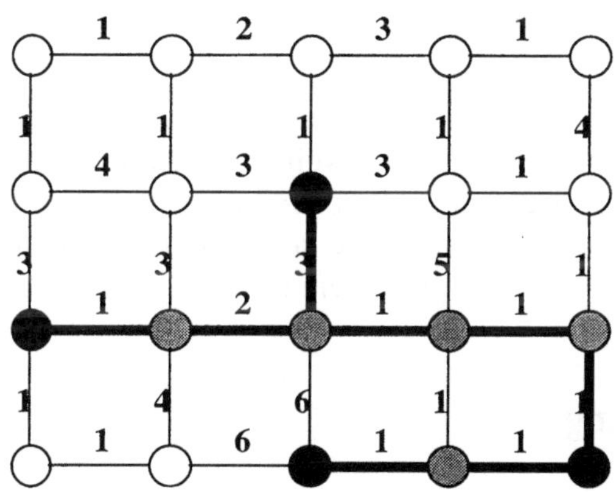

(b) Minimum Spanning Tree Using Steiner Nodes

Figure 9. GSMT Problem

The problem is defined as:

`Problem: Given a weighted graph` $G=(V,E)$ `, and a set of nodes` $N \subseteq V$ `, find a minimum-cost tree` $T=(V',E')$ `with` $N \subseteq V' \subseteq V$ `and` $E' \subseteq E$ `.`

Each graph edge e_{ij} has a weight w_{ij}, and the cost of a tree or any subgraph is the sum of the weights of its edges. Here, N is the set of demand points, and $V-N$ are called the Steiner points. In the FPGA routing problem, the set of nodes, N , refer to the pins of the logic blocks that have to be connected together utilizing the Steiner nodes that represent the routing resources. Fig. 9 illustrates the GSMT problem.

The flow is as shown in Fig. 8, and is similar to that given in [Alexander94][Kahng95]. The target architecture description, placement of logic blocks, and connectivity of logic blocks are the inputs to the tool. Since the routing is sequential in nature, the nets are ordered to improve the chances of successful routing. The core of the tool is a GSMT algorithm [Alexander96], which attempts to route the net on the target graph. If not all of the nets can be routed, the nets are reordered, and another pass is made. Upon successful routing, the information regarding the routes for all of the nets is available for further analysis.

7.1 Input

The inputs to the router are the weighted graph representation of the target architecture, the location of the logic nodes on this graph, and the connections required between the logic nodes. The location of the logic nodes is obtained from the placement step.

7.2 Net Ordering

The GSMT algorithm attempts to route each net using available routing resources. Due to the sequential nature of routing the nets, the nets that have already been routed will affect the resources available for routing nets lower down in the order. This means that the ordering of the nets can have an impact on the final routing of the netlist.

A net with a smaller bounding box has fewer possible routes. The bounding box is defined by the location of the pins connected by the net. Based on this, a simple heuristic is used for ordering the nets. The nets are ordered in order of increasing bounding box size, so that nets with fewer choices are routed first.

During an unsuccessful routing pass, the nets that cannot be routed are tagged. The nets are reordered with the tagged nets moved to the top of the order. This will ensure that the nets which could not be routed have access to more routing resources during a successive pass, and hence a higher chance of being routed.

7.3 GSMT Algorithm

The GSMT problem is known to be NP-complete [Hwang92]. There exist a number of heuristics to tackle this problem. The heuristic of Kou, Markowsky and Berman (KMB) [Kou81] solves the GSMT problem in polynomial time. This heuristic has a performance bound of $2 \cdot \left(1 - \frac{1}{L}\right)$, where L is the maximum number of leaves in any optimal solution. The IKMB heuristic [Alexander96], which uses iterated heuristics, is used as the graph routing algorithm.

7.3.1 KMB Heuristic [Kou81]

The KMB algorithm is given below

Input: A graph $G = (V, E)$ with edge weights w_{ij} and a set $N \subseteq V$

Output: A low-cost tree $T = (V', E')$ spanning N (i.e. $N \subseteq V'$ and $E' \subseteq E$)

$G' = (N, N \times N)$, with edge weights $w'_{ij} = dist_G(n_i, n_j)$

Compute $T = (N, E'') = MST(G')$

$G'' = \bigcup_{e_{ij} \in E''} path_G(n_i, n_j)$

Compute $T' = MST(G'')$

Delete pendant edges from T' until leaf nodes are in N

Output T'

The steps are illustrated in Fig. 10.

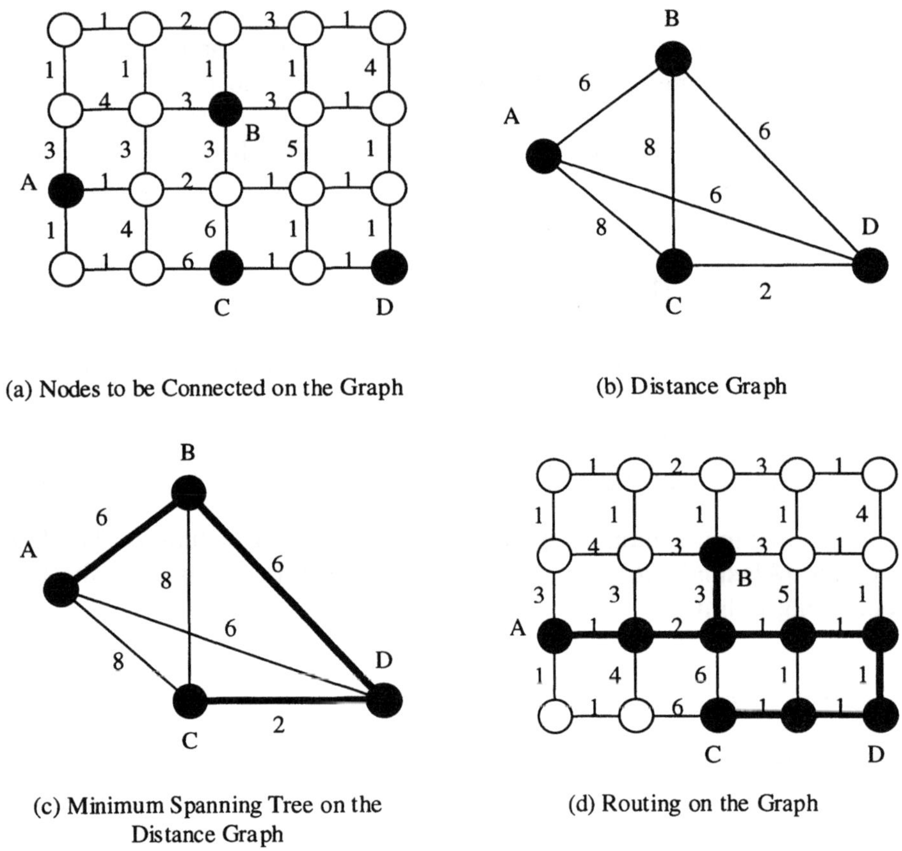

(a) Nodes to be Connected on the Graph

(b) Distance Graph

(c) Minimum Spanning Tree on the Distance Graph

(d) Routing on the Graph

Figure 10. KMB Algorithm

- Step 1: Construct the distance graph G' over N. The weight of each edge e'_{ij} is given by w'_{ij} which is the cost of the shortest path in G between n_i and n_j. This shortest path is denoted by $path(n_i, n_j)$.
- Step 2: Compute the minimum spanning tree of G', $MST(G')$.

- Step 3: Expand each edge e'_{ij} into the corresponding path $path(n_i, n_j)$, to yield G'' that spans N.
- Step 4: Compute the minimum spanning tree of G'', $MST(G'')$, and delete pendant edges from $MST(G'')$ until all the leaves are members of N.

7.3.2 Iterated KMB (IKMB) [Alexander96]

The iterated heuristic tries to improve the solution obtained using an existing Steiner algorithm by greedily selecting Steiner nodes which reduce the cost with respect to the original solution. The definition of IKMB is given by:

Given a set of Steiner candidate nodes $S \subseteq V - N$, the cost savings of S with respect to KMB is defined as
$\Delta KMB(G, N, S) = \cos t(KMB(G, N)) - \cos t(KMB(G, N \cup S))$.

The algorithm is described below:

Input: A graph $G = (V, E)$ with edge weights w_{ij} and a net $N \subseteq V$
Output: A low-cost tree $T' = (V', E')$ spanning N (i.e. $N \subseteq V' \subseteq V$ and $E' \subseteq E$)
$S = 0$
Do Forever
  $T = \{t \in V - N \mid \Delta KMB(G, N, S \cup \{t\}) > 0\}$
  If $T = 0$ **Then Return** $KMB(G, N \cup S)$
  Find $t \in T$ with maximum $\Delta KMB(G, N, S \cup \{t\})$
  $S = S \cup \{t\}$

The heuristic starts with an empty set of candidate Steiner nodes, $S = 0$. Then each node in $V - \{N \cup S\}$ is tried until a node t, which maximizes $\Delta KMB(G, N, S \cup \{t\})$, is found. This procedure is repeated with $S \leftarrow S \cup \{t\}$. The cost of KMB to span $N \cup S$ will decrease with each added t. The routine terminates when there is no $t \in (V - N) - S$ such that $\Delta KMB(G, N, S \cup \{t\}) > 0$.

The IKMB heuristic is illustrated in Fig. 11. For the problem in Fig. 11(a), the solution using KMB is shown in Fig. 11(b) with a cost of seven. But it is seen that following the IKMB heuristic, Steiner node S3 can be added to reduce the spanning cost to 6 as shown in Fig. 11(c). The final solution is shown in Fig. 11(d).

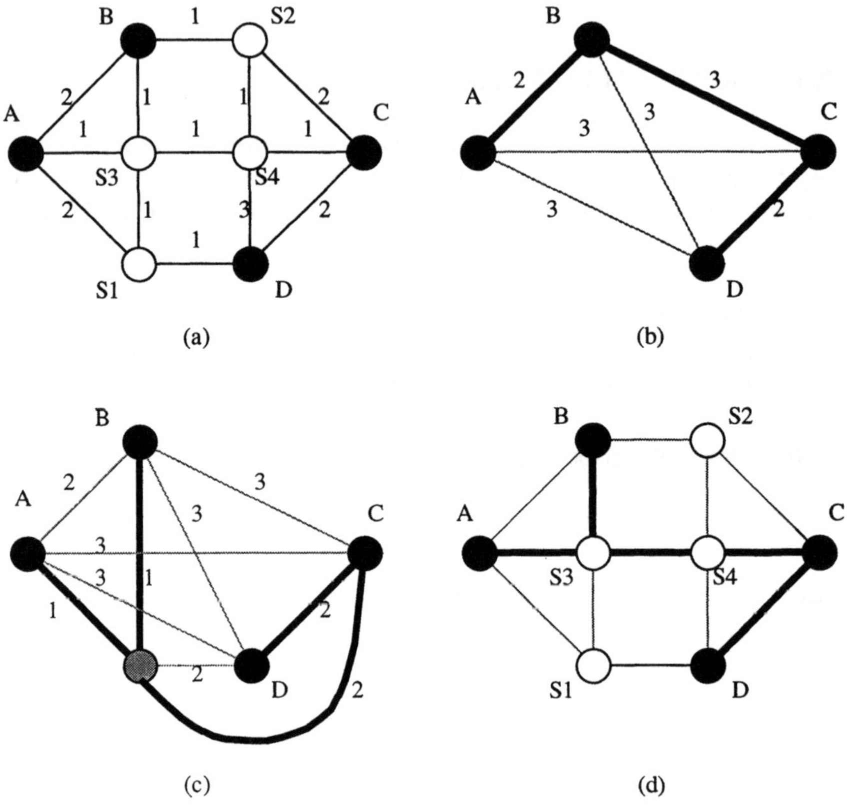

Figure 11. IKMB Heuristic

The performance bound of the IKMB heuristic is no worse than that of the KMB heuristic it uses, since if no improving Steiner node is found, the output is identical to that of the KMB heuristic. The KMB heuristic is used inside the IKMB algorithm, and thus inherits the performance bound of ≤2 times optimal.

The time complexity of the algorithm in the worst case is $O(|V| \cdot t(KMB))$ per iteration, where $t(KMB)$ is the time complexity of the KMB heuristic. This can be improved significantly by extracting the common one-time computations in the KMB heuristic. The shortest path computation, which is required in the KMB heuristic falls into this category.

7.3.3 Congestion Management

The chance of a successful routing is improved if the routed nets are uniformly distributed over the routing resources, rather than crowded in certain locations. This is especially true of long nets, which traverse a significant fraction of the array.

The possible reasons for this crowding are:

- A particular path might be offering a lower cost. Usually the cost improvement is minimal over another route.
- The order in which the interconnect resources are examined can result in tracks from a particular channel being selected over identical cost tracks occurring in other channels.

One way to overcome the congestion problem is by artificially increasing the cost of routes through congested routing channels. This will force the router to seek alternate, cheaper routes.

To implement this, tracks are grouped into specific routing groups based on proximity. For the example in Fig. 12, all tracks in one vertical channel can be grouped together, similarly for a horizontal channel. When a track in a particular routing group is used, the cost of all the other available tracks in the routing group is increased. As the number of available tracks in the channel reduces, the cost of the available tracks increases. This will ensure that the routes through congested routing channels become more expensive, and the routes will be distributed to other routing channels with lower congestion. This is illustrated in Fig. 12(b). The grouping information is included in the target architecture description, which is supplied to the router.

This method does have the disadvantage that the grouping of the resources has to be done before routing by the designer. At present, the grouping is done based on the physical proximity of the resources that are interchangeable.

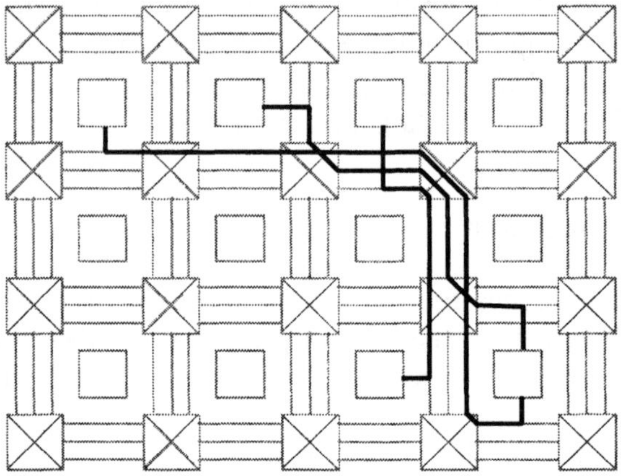

(a) Congested Routing

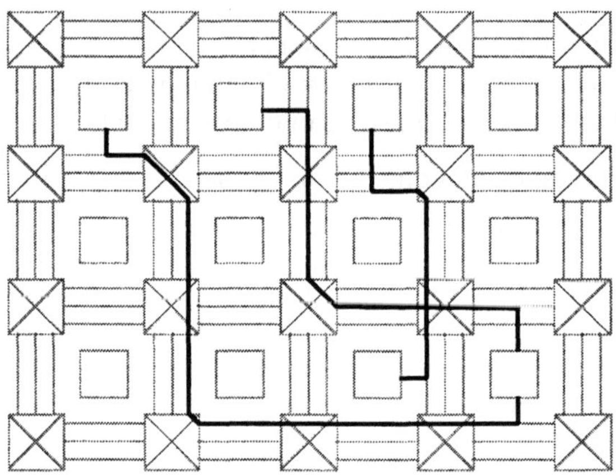

(b) Distributed Routing

Figure 12. Congestion Management

7.3.4 Bounding Window

The time complexity of the KMB heuristic, and that of the IKMB wrapper, is dependent on the graph size. Reducing the size of the graph can reduce the run time. The search space, and hence the size of the graph, can be reduced by using a bounding box. The bounding box limits the possible routes that are evaluated for routing a specific connection. The location of the pins belonging to the net is taken into account in defining the bounding box.

Constraining the number of available routes can result in unsuccessful routing of the net. To solve this problem, an adaptive bounding box is used. If a routing pass is unsuccessful, during the next pass the size of bounding box is increased for the failed nets. This ensures that the nets have more routes to choose from, and increases the probability of a successful routing.

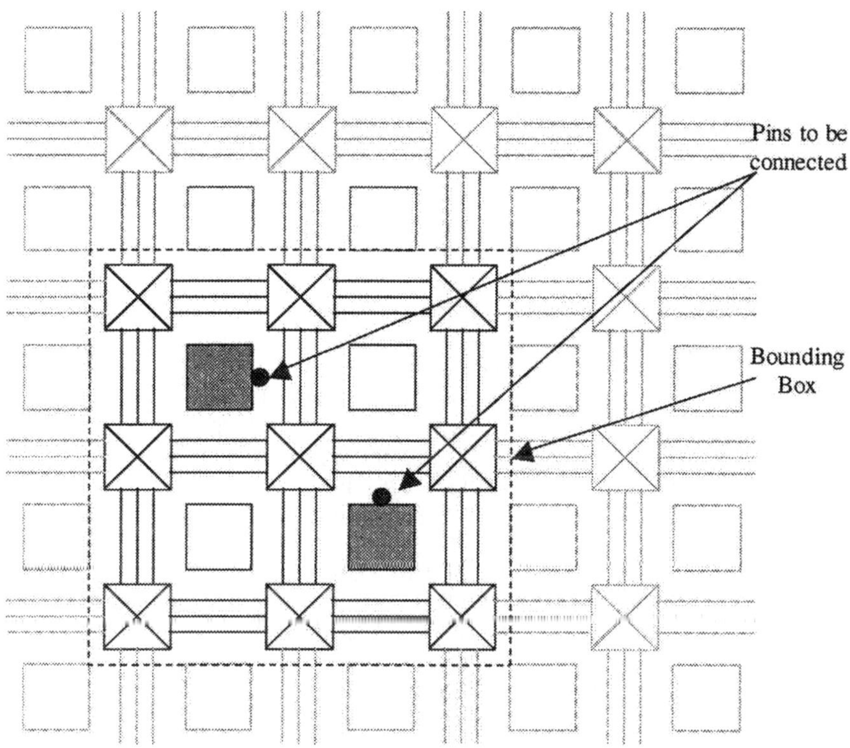

Figure 13. Bounding Box to Limit Graph Size

For this option in the router to work, the logic and routing resources have to be annotated with relative position information. In the flow implemented, the designer provides this information as part of the description of the graph. The size of the bounding box with respect to the position of the pins is also user-defined.

8 EXTRACTION

The output of the placement and routing flow is the route taken by each source-sink pair in each net. A sample output is shown in Fig. 14. The Steiner nodes used to route each source-sink pair correspond to the routing resources defined in the architectural description.

```
0 2484 2492 2508 2457 2443 2435
27 1674 1682 1698 1647 1633 1625
20 756 764 780 1215 1201 1193
17 1028 1036 1051 1487 1473 1465
15 2864 2872 2887 2837 2823 2815
28 1136 1144 1159 1595 1581 1573
31 1622 1630 1645 1217 1203 1195
44 1568 1576 1591 2027 2013 2005
39 1460 1466 1482 1917 1903 1895
36 542 548 564 999 985 977
35 1192 1200 1208 1186 745 1155 1147 1139
14 3294 3303 3311 2905 2914 2884 2876 2868
23 1242 1250 1258 1239 1210 1202 1194
10 2916 2924 2940 2889 2875 2867
8 3240 3250 3258 3292 2860 2830 2822 2814
7 2754 2763 2771 2365 2775 2767 2759
7 2700 2708 2724 2781 2767 2759
7 2808 2816 2832 2781 2767 2759
9 2758 2766 2774 2745 2776 2768 2760
9 2701 2710 2718 2745 2776 2768 2760
9 2811 2819 2834 2782 2768 2760
```

Net ID
Sink Node
Source Node

Figure 14. Sample Output of the Placement and Routing Flow

The above information, along with the cost of each resource, is sufficient to extract the delay information of each source-sink path and the overall energy cost of each electrical net. The delay is calculated using the Elmore delay model [Elmore48]. The energy cost is obtained from the total capacitance of each net.

9 CONCLUSION

The evaluation of FPGA architecture requires a complex flow involving logic synthesis, placement, routing, and extraction of the implementation on the target architecture. Most of the existing tools are tied tightly to specific architectures and are not amenable to large modifications to the architecture.

In this work, a flexible implementation flow is developed to explore different architectural structures. The architecture is described as a weighted graph structure that encapsulates all of the required information. The nodes in the graph correspond to the logic blocks and the routing resources. The edges between the nodes represent the connectivity possible between the logic and routing resources. The weights on the edges correspond to the capacitive cost, and hence the energy of using the resources.

Existing algorithms and tools are modified to operate on this description. MIS-FPGA is used to synthesize the applications onto a netlist of lookup tables. A traditional simulated annealing based algorithm is used for placement. Modifications are done during the candidate selection process to improve the quality of the placement.

The routing is posed as a GSMT problem. There are numerous algorithms that can be used to solve the MST problem. The IKMB algorithm is chosen because of the well-defined performance bound of $2 \cdot \left(1 - \frac{1}{L}\right)$, where L is the number of leaves in the graph. The run time of the algorithm is improved by constraining the search space. The router also does congestion management based on grouping properties provided with the architecture description.

One of the features of this router is that different GSMT algorithms can be used instead of the IKMB algorithm. For example, in this work, the cost function used is the capacitive cost. This requires minimizing the cost of the routing resources used to route the entire net, without consideration for any specific source-to-sink path. It is conceivable that for different applications, the dominant concern might be the timing path. In such a situation it will be appropriate to replace the IKMB algorithm with another algorithm more suited to solve the problem at hand. For example, the graph Steiner arborescence heuristic based on a path-folding strategy [Alexander96] can be used to generate a shortest-paths tree to yield the greatest possible wirelength savings while maintaining the shortest-paths property.

The flow described in this chapter is used in the rest of the work to evaluate the different architectural features. Note, however, that the flexibility of the flow is obtained at the cost of the run-time as compared to an architecture specific placement and routing flow.

Chapter 4

LOGIC AND INTERCONNECT ARCHITECTURE

1 INTRODUCTION

The energy and speed performance of the commercial FPGA architecture are dominated by resistance and capacitance contributions from the routing switches. Increasing the connectivity in the interconnect architecture usually involves an increase in the number of routing switches. Increasing the number of routing switches can improve the flexibility of the architecture at the price of area, speed, and energy. The architectural optimization process is to evaluate the trade-off between the flexibility of the architecture and the performance metrics. It is also an effort at evaluating smarter ways of distributing the routing switches to improve the performance metrics without sacrificing flexibility. The performance metric chosen for evaluating the architecture is Energy-Delay Product (EDP). This ensures that energy efficiency is not obtained at the cost of speed performance.

An exhaustive evaluation of the entire architecture design space will be difficult. A more practical approach is to take a proven existing architecture as a starting point. The software environment described in Chapter 3 is used to implement the benchmark suite on the target architecture, and extract the performance data.

2 RELATED RESEARCH

There has been considerable work in the area of FPGA architecture. The work can be broadly divided into two categories: logic block structures and interconnect architectures.

The studies on the structure of the logic block were targeted at understanding the relationship between the functionality of the logic block and the area of the FPGA. The work by Rose, et al. looked at the effect of the logic block functionality on area efficiency [Rose90a]. Results indicated that the best number of inputs to use was between three and four. The study also indicated that a D-flip-flop should be included in the logic block. The study into multiple output LUTs [Kouloheris91][Kouloheris92] showed that a 4-input LUT gives the minimum area. Looking at the relationship between logic block architecture and the speed of the FPGA [Singh92], it was shown that a 5 or 6-input LUT attained the lowest delays. This was attributed to the reduction in number of stages of slow programmable routing.

Based on the above results, the use of heterogeneous logic blocks has been advocated to provide a better tradeoff between speed and density [He93]. For example, a combination of 6 and 4-input LUTs was predicted to have the same area as a homogenous 4-input LUT FPGA, while improving the speed by 25%.

Chung, et al. studied the use of hierarchical logic blocks to improve the speed of FPGAs. In this topology, basic logic blocks are chained together using hard-wired connections rather than programmable connections [Chung91]. The work looked at the tradeoff between the resulting area penalty, and the speed gain. It was shown that ~26% reduction in delay can be obtained at the cost of a 5% increase in area.

The work by Betz, et al. looked at the area efficiency of cluster-based logic blocks [Betz97a]. The logic block is realized using a cluster of 4-input LUTs. It was shown that a cluster size of four can achieve an area reduction of 5 to 10% when compared to a logic block with one 4-input LUT.

The Symmetric Mesh architecture has been one of the most popular interconnect structures. Most of the studies on logic block structures assumed a mesh interconnect structure. The work by Rose, et al. explored the relationship between the routability of an FPGA and the flexibility of its interconnect structure [Rose90]. It was shown that the minimum number of switches, and hence the minimum area, is obtained for a switch box flexibility between three and four, and a connection box flexibility between 0.7 and 0.9.

Variations in the interconnect structure in terms of segmented architectures were explored, and shown to improve the speed performance [Brown96]. Segmented routing has also been proposed to improve the predictability of routing [Ochotta98]. The study by Betz, et al. looked at the distribution of channel size over the array [Betz96].

While the above results were obtained using empirical methods, there has been work done to develop stochastic models to predict the routability of the Mesh architecture [Brown93]. By modeling the channel density with a Poisson distribution, an analytic expression was derived to predict the routability of the circuit in the FPGA. The predicted values matched experimental results.

Aggarwal, et al. showed that hierarchical architectures obtain significant reduction in the switch count as compared to Symmetric Mesh architectures [Aggarwal94]. A study by Chan, et al. into the area-speed tradeoff for the hierarchical architecture [Chan96] shows up to a 64% improvement in speed.

Lai, et al. developed an analytic model to evaluate minimum switch populations required for implementing different *m*-ary hierarchical structures [Lai98]. It is shown that a 4-ary tree is optimal from the perspective of switch count. An area reduction of 40% is predicted as compared to Mesh architecture.

Hierarchical architectures have been used to push the performance of FPGA architectures. Tsu, et al. demonstrated an FPGA operating at 250MHz by using a 2-ary tree structure [Tsu99]. The architecture also uses pipelining at the interconnect level to achieve higher speed.

These works look at the FPGA from an area and speed perspective. The issues involved in designing a low-energy FPGA have not been evaluated.

3 ENERGY-DELAY COMPONENTS

The energy and delay of the FPGA are dominated by the interconnect architecture, followed by the clock distribution network, as shown in Chapter 2. Based on this observation, architectural modifications are aimed at reducing the overhead of interconnect and clock distribution structures to the total energy.

Fig. 1(a) shows a connection realized in an FPGA interconnect fabric, by enabling routing switches in the connection and switch boxes. The switch level realization of the connection and switch boxes is shown in Fig. 1(b). The connection box is realized in the form of a pass-transistor structure to connect one pin from the logic block to the routing channel. The switch level circuit of the switch box shows the connections between corresponding tracks in the routing channels.

The parasitic contribution from the switches and the metal trace constitute the total resistive and capacitive components of the interconnect. The switches are realized using NMOS pass transistors with the control signal applied to the gate terminal.

The equivalent parasitic model of the switch is shown in Fig. 2. The main source of the capacitive component is from the reverse-biased source-bulk and drain-bulk *pn*-junctions. R_{on} gives the equivalent resistance of the pass transistor when it is in the routing path.

The parasitic model of the metal trace is a distributed *RC* network. The resistance is from the sheet resistance of the metal, and the capacitance is from the area and fringe capacitance. In the present deep sub-micron design regime the fringe capacitance is the dominant component for typical wire widths.

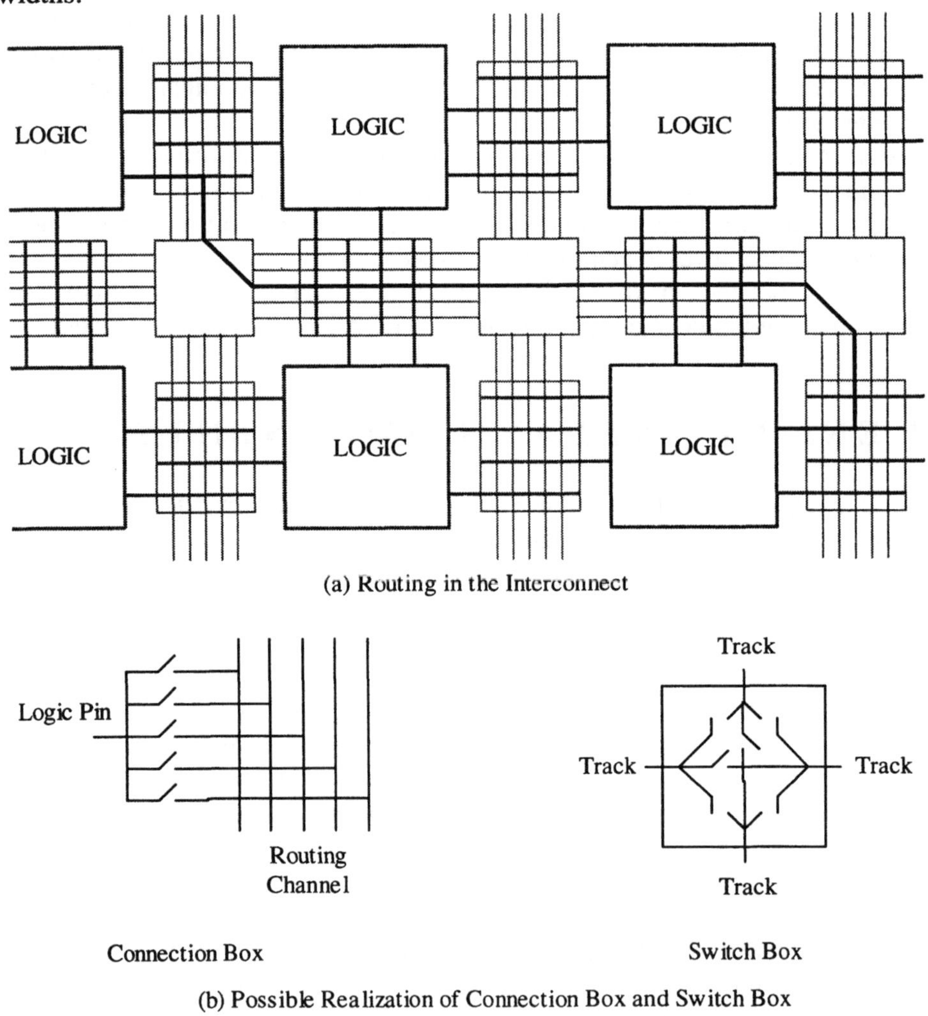

Figure 1. Routing in a Mesh Fabric

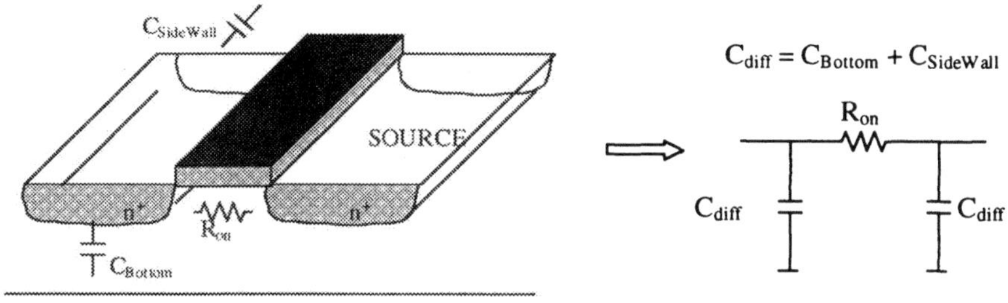

(a) Parasitics from the Routing Switch

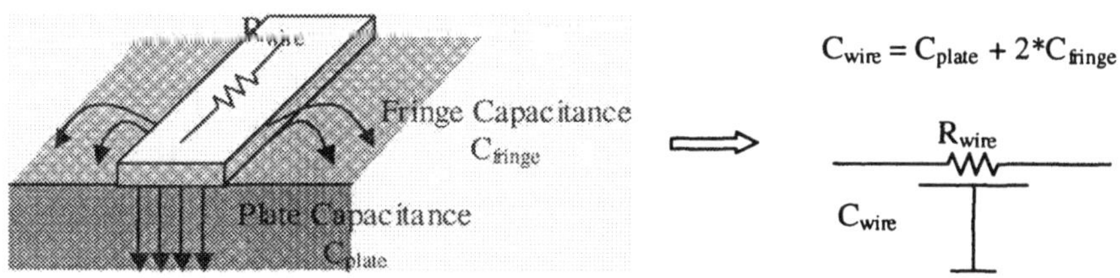

(b) Parasitics of the Metal Trace

Figure 2. Parasitic Components in the Interconnect

Based on the switch and wire parasitic, the interconnect route in Fig. 1(a) can be represented with an *RC* network. For typical parasitic values, $R_{wire} \ll R_{on}$, and can be neglected. Fig. 3 gives the equivalent *RC* network. The energy can be calculated from the capacitance of the route, while the delay is dependent on the resistance and capacitance. For example, the capacitance of a routing segment is be given by,

$$C_{seg} = 10 \cdot C_{diff} + C_{wire}$$

This can be used to model the energy of the route as,

$$Energy(E) \propto 50 \cdot C_{diff} + 4 \cdot C_{wire}$$

The delay of the route can be computed using the Elmore delay model [Elmore48], to give,

$$Delay(D) \propto 10 \cdot R_{on} \cdot C_{wire} + 125 \cdot R_{on} \cdot C_{diff}$$

This method of modeling the interconnect is used to compute the cost of the architectural modifications.

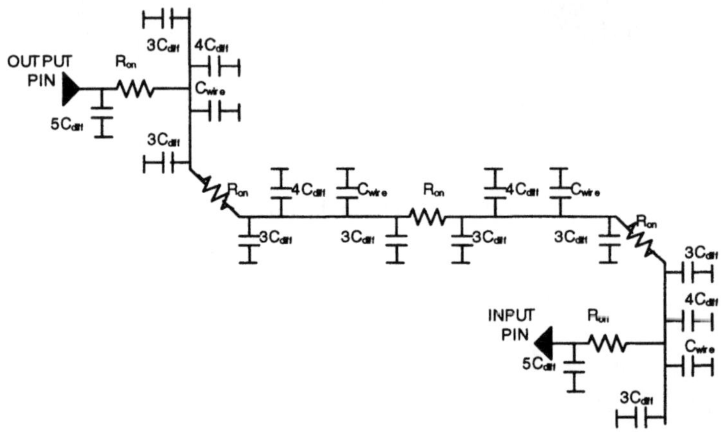

Figure 3. Interconnect Routing and RC Equivalent

4 ARCHITECTURAL COMPONENTS

The Symmetric Mesh architecture is a common standard architecture, and variations of this basic structure exist in most of the commercial structures. This structure has been researched extensively. The Symmetric Mesh architecture is chosen as the starting point. The main architectural components are the logic block, interconnect architecture and the clock network.

It was shown earlier that the logic block is not interesting from its inherent energy. The construction of the logic block is interesting from the effect it has on the usage of interconnect resources. If the logic block is inefficient, then more logic blocks, and hence more interconnect, will be used to implement a given function. This will increase the contribution of the interconnect to the total energy. The logic block is defined by the granularity of the logic block and the internal connections in the block.

The interconnect is defined by multiple parameters and realization options. It is practically impossible to do an exhaustive analysis of all the possible options. One way of tackling this problem is to analyze existing structures to have a better understanding of the importance of different parameters.

The clock network is the next major contributor to the total power. Since all the logic blocks in the array have flip-flops associated with them, the clock has to be distributed throughout the chip. Energy efficient methods to distribute the clock have to be investigated to exploit the regular and symmetric nature of the distribution.

5 LOGIC BLOCK

The logic block is defined by its internal structure and the granularity. The structure defines the different kinds of logic that can be implemented in the block, while the granularity defines the size of the function that can be implemented.

5.1 Logic Block Structure

The logic block is chosen to be lookup table based. The advantage of using a k-input LUT (k-LUT) is that it can realize any combinational logic with k inputs. Before exploring the granularity of the logic block, the internal structure has to be analyzed. Previous work that evaluated the effect of the logic block on the FPGA architecture [Rose90a] used a k-input LUT with a single output as the logic block. This structure is better for implementing random logic functions than for datapath-like bit-slice operations. This can be illustrated using examples shown in Fig. 4 and Fig. 5.

Fig. 4 shows a random combinational logic. If a single output logic block structure were to be used, increase in the LUT size could easily be used to reduce the number of logic blocks required. Fig. 4 illustrates how two 5-input LUTs can be used to cover the same logic as four 3-input LUTs.

Arithmetic operations cannot take advantage of the increased LUT size as easily as random logic. For example, consider the 2-bit adder shown in Fig. 5. The smallest component is a 3-input operation. Increasing the block granularity will not improve the block count, since the increased functionality is useless without a corresponding change in the number of outputs. To perform an unbiased evaluation, the structure should be such that any increase in logic block granularity in terms of LUT size is equally useful for both random logic and datapath operations.

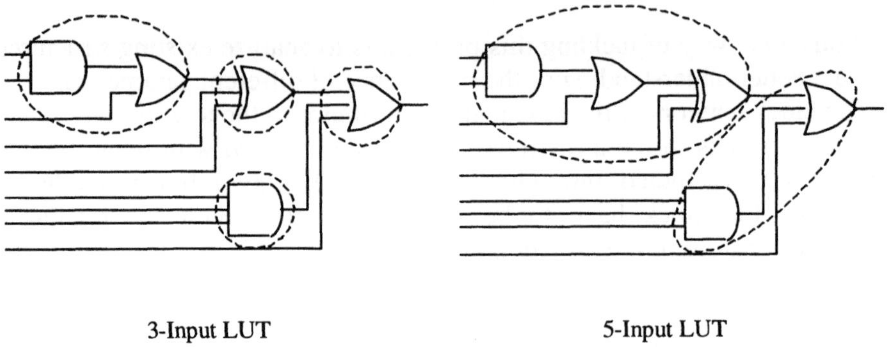

Figure 4. Random Logic and LUT size

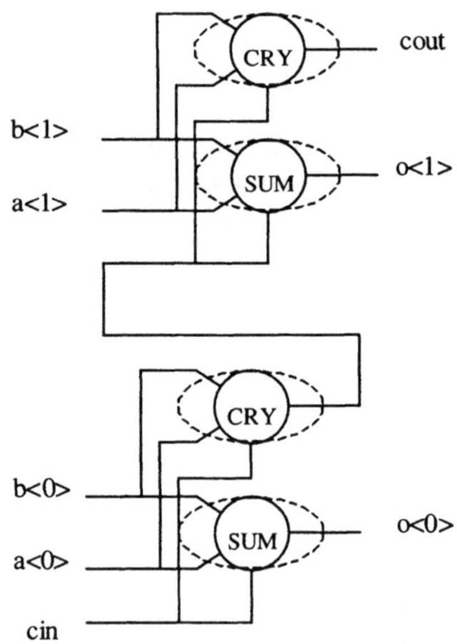

Figure 5. Arithmetic Block

Fig. 6 shows a possible logic block structure that can be used for implementing both random and arithmetic logic without any wastage of logic resources. The logic block is made up of a cluster of 3-input LUTs that can be connected up to realize a larger function. The number of output pins is also increased along with the logic block size to facilitate its use as an arithmetic unit.

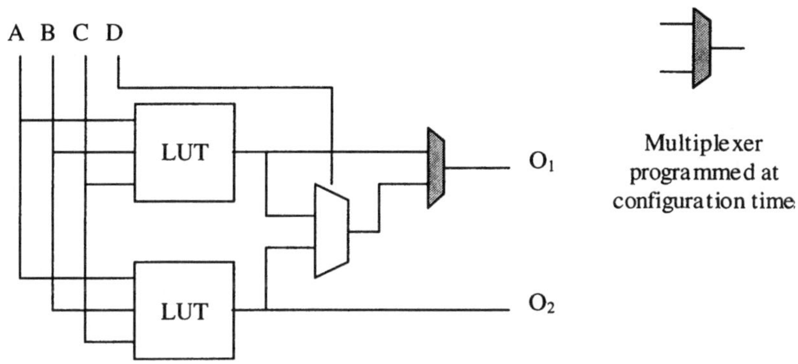

(a) 4-Input LUT / 1-Bit Arithmetic Operator

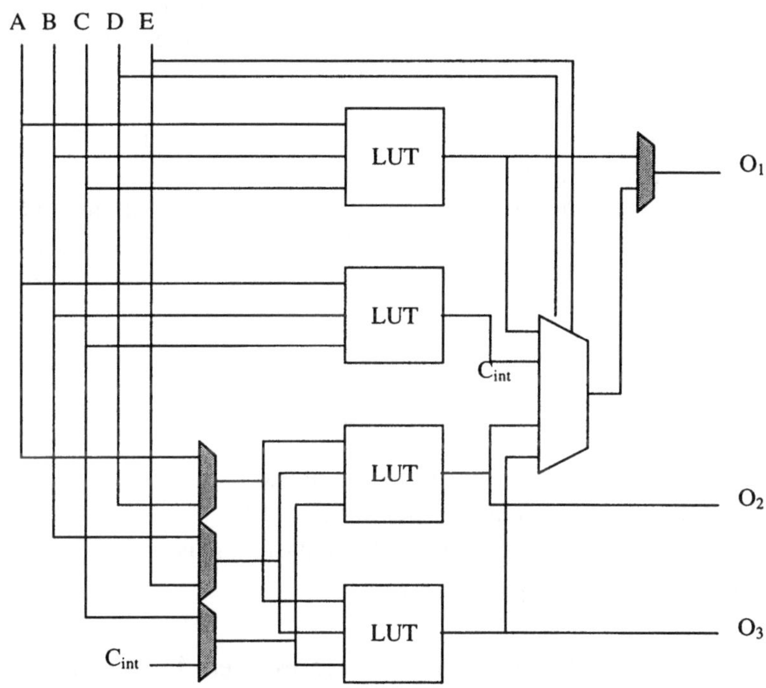

(b) 5-Input LUT / 2-Bit Arithmetic Operator

Figure 6. Different Logic Blocks Realized Using Cluster of 3-Input LUTs

5.2 Optimal Logic Block Granularity

The implementation flow described in Chapter 3 is used to evaluate the impact of the logic block granularity on the energy performance of the array. For each logic block granularity, MIS-FPGA is used to synthesize the application to a netlist of LUTs. The total capacitance of the routed nets is used as a measure of the cost of implementing a function. At the conclusion of the routing step, complete information of the routes taken by all the nets is available. This information, combined with the cost of each of the routing segments, can be used to compute the total cost of routing the nets.

The granularity of the logic block is described in terms of the equivalent lookup table size. The logic blocks are made up of cluster of 3-input LUTs, with the structure described in the previous section. Symmetric Mesh architecture as described in Chapter 1 is used. The connection boxes have full connectivity, i.e. the pins connect to all the tracks in their respective routing channels. A disjoint switch box with a flexibility of three is used.

As the granularity of the logic block is increased, the number of logic blocks required to realize each function reduces. This is accompanied by a reduction in the number of nets routed in the routing channels. The capacitance of the routing segment is dependent on the number of logic block pins accessing the track, and the contribution of the switch box.

The switches in the connection box and the switch box are assumed to be of the same size. The pins of the logic block are uniformly distributed on all sides of the logic block. The capacitance of the routing segment is given by,

$$C_{seg} = \frac{(M+N)}{2} \cdot C_{diff} + 2 \cdot F_s \cdot C_{diff}$$

Where,

- M is the number of input pins per logic block
- N is the number of output pins per logic block
- C_{diff} is the capacitance contribution of one pass transistor
- F_s is the flexibility of the switch box

As the granularity of the logic block increases, there is an increase in the number of input and output pins. This has a direct impact on the capacitance of the routing segments. The cost of the each routing segment as a function of the logic block granularity is given in Table 1.

Table 1. Capacitance of Routing Segment as a Function of Logic Block Granularity

Granularity (k-LUT)	# Input Pins	# Output Pins	Capacitance (C_{seg})
3	3	1	8
4	4	2	9
5	5	3	10
6	9	5	13

The experimental results over a set of benchmarks is given Table 2 and plotted in Fig. 7. As the size of the logic block is increased from a 3-input LUT to a 5-input LUT, the total cost of the interconnect reduces. This can be attributed to the fact that as the capacity of each block is increased, more logic can be packed into each logic block. This reduces the number of nets routed on the expensive programmable interconnect resource. Hence, even though the cost of each routing segment increases, the reduction in the usage of the resources results in a decrease of the total cost of the interconnect.

However, as the granularity is increased beyond that of a 5-input LUT, the interconnect cost increases. This is the result of two factors: under-utilization of large granularity logic blocks and higher cost of the routing segments due to the larger number of pins per logic block.

For the random logic functions, a 4-input or 5-input logic block gives the minimum energy. The datapath functions have a strong bias for a 5-input logic block. Therefore, granularity equivalent to that of a 5-input LUT is found to be optimal from an energy perspective. It should be noted that the logic block is comprised of a cluster of 3-input LUTs to facilitate efficient implementation of arithmetic operations. The final logic block can implement a 5-input random logic function or a 2-bit arithmetic function. All of the outputs can be registered.

This experiment does not take into consideration the cost of the logic resource. The logic blocks used in this experiment are of small enough granularity that the cost of the interconnect dominates the total cost. So, the cost of the logic block can be ignored. If the experiment is repeated for larger granularities, this assumption will have to be analyzed.

Table 2. Routing Statistics for the Benchmark Suite

Benchmark	Logic Block Size	# Logic Blocks	#Routing Segments	Total Capacitance	Normalized Capacitance
s526	3	133	734	5872	1.31
	4	93	595	5355	1.19
	5	69	449	4490	1.00
	6	64	460	5980	1.33
term1	3	475	3070	24560	1.43
	4	310	2270	20430	1.19
	5	245	1826	18260	1.06
	6	180	1323	17199	1.00
ACS	3	148	959	7672	1.13
	4	116	754	6786	1.00
	5	60	702	7020	1.03
	6	31	557	7241	1.07
correlator	3	166	760	6080	1.44
	4	89	470	4230	1.00
	5	61	473	4730	1.12
	6	38	466	6058	1.43
BMA	3	400	2219	17752	1.54
	4	210	1375	12375	1.08
	5	130	1151	11510	1.00
	6	65	980	12740	1.11
FIR	3	111	660	5280	1.73
	4	65	401	3609	1.18
	5	33	306	3060	1.00
	6	17	315	4095	1.34
IIR	3	315	1888	15104	1.54
	4	172	1093	9837	1.00
	5	91	1005	10050	1.02
	6	48	898	11674	1.19

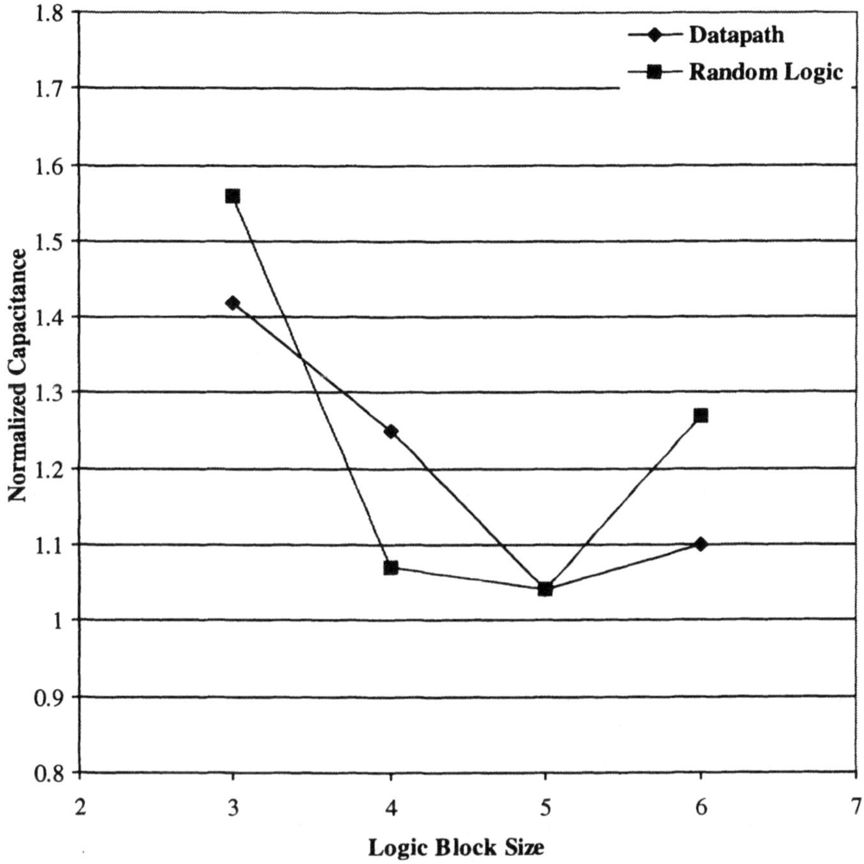

Figure 7. Interconnect Capacitance as a Function of Logic Block Size

6 GOAL OF INTERCONNECT OPTIMIZATION

The cost of the interconnect can be reduced by reducing the number of switches or by reducing the size of the switches. The number of routing switches is determined by the connectivity in the architecture. The size of the switches is dependent on the average length of the routes, and the speed performance that has to be supported.

6.1 Flexibility and Energy

Reducing the flexibility of the connection and switch boxes is equivalent to reducing the number of switches connected to the routing segments. This will reduce the capacitive load on the routing segments, and reduce the total energy. Reducing the flexibility can result in more routing segments being required to realize a given connection. Hence, this method can be counterproductive if the total capacitance of the route is greater than the capacitance before the modifications to the architecture. In the extreme case, reduction in flexibility can result in unsuccessful routing of the application. Hence, any flexibility modification has to be accompanied by an architecture evaluation to ensure that applications can still be efficiently implemented.

6.2 Average Path Length and Energy

The delay of a route in the FPGA interconnect has a direct dependence on the size of the routing switches. Increase in switch size reduces the series resistance, but will increase the diffusion capacitances. The size of the switch is determined by the system frequency that needs to be supported, and the average path length. The average path length is a rough estimation of the length of most of the routes. In the Symmetric Mesh architecture, this will be in terms of the number of single segments, while in a hierarchical structure the path length is in terms of the number of levels in the tree structure. The switches have to be designed so that the system frequency could be supported if all the routes are of average path length.

One of the disadvantages of designing for the average path length is that the routes which are shorter than the average path length are penalized because of the wider switches. To illustrate this, an interconnect structure as shown in Fig. 1 is assumed. For a path length, i, the switch sizes are optimized to obtain a 1.5nS delay in a 0.25μm process technology. The optimal switch size for each of the path length, i, is w_i. Obviously, the longer paths require wider switches to meet the delay specifications. A switch of size w_i cannot meet the delay specification for any path length $j > i$, but it can meet the specification for any path length $j \le i$. Fig. 8 shows the energy consumptions of the different switch sizes for different path lengths.

It is seen that the penalty paid for using a switch size intended for another path length is quite high. For example, if the switch size is optimized for a path length of seven, then the penalty paid for routing a path length of three is approximately five times. This underlines the interdependence between path length, speed performance and energy dissipation in the interconnect architecture. This is exploited in the next section during the exploration of the interconnect structure.

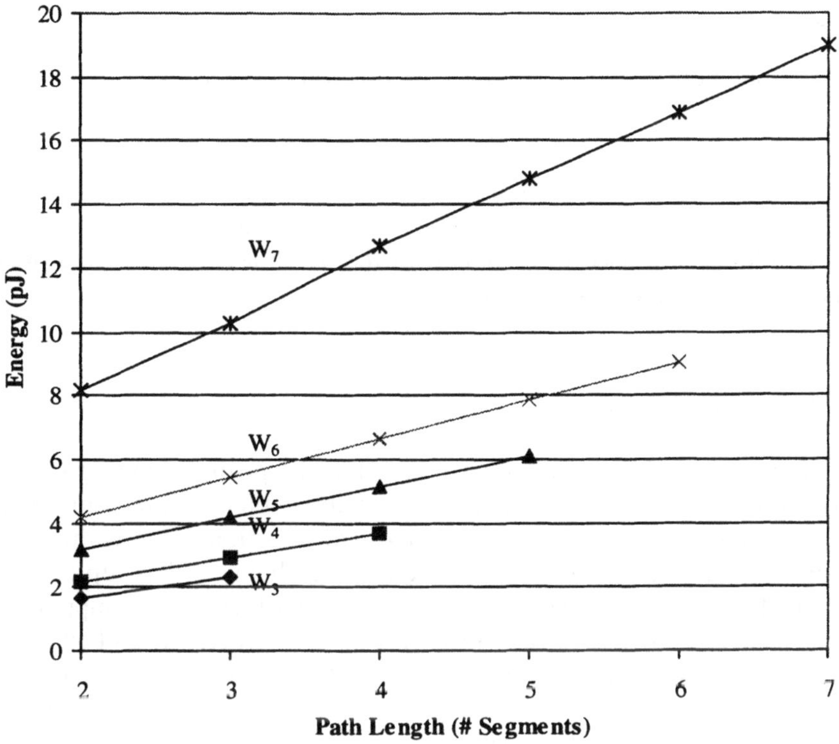

Figure 8. Path Length, Switch Size, and Energy

7 INTERCONNECT ARCHITECTURE

Using only the Mesh structure results in a simple interconnect architecture, where the same routing structure is used independent of the length of the routes. The analysis of the relationship between the width of the switch, path length, delay and energy discussed in the previous section indicates that this simplicity is achieved at the cost of energy. The results from the previous section indicate that energy improvements can be obtained by having different levels of interconnect to realize different classes of path lengths.

In this work the routes have been broadly classified into three regions:

- Short nets that realize connections between adjacent logic blocks.
- Intermediate length nets.
- Long nets that span a significant fraction of the array.

The interconnect architecture is divided into three levels, with each level targeted at a specific class of nets. The Mesh architecture is used as the backbone, and used to provide connections for intermediate length wires. The other two levels are used to complement the Mesh architecture, and provide connections for short nets and long nets.

7.1 Mesh Architecture

The Symmetric Mesh architecture is used as the primary routing structure. It consists of a two-dimensional array of logic blocks interconnected by routing channels in the vertical and horizontal directions. The width, *W*, is a measure of the number of tracks in the routing channel. This represents the number of simultaneous independent connections that can be supported by the channel. The structure is as shown in Fig. 9. The connection box controls the connection between the input/output terminals of the logic block and the routing channels. The switch box determines the connections possible between the different routing channels.

Previous work evaluating the effect of the connection box and the switch box on the routability [Rose90b] indicates switch box flexibility, F_s, of three, and a high connection box flexibility, F_c, of 0.8. If F_c is less than 1.0, then the distribution of the switches in the connection box is important, and can affect the routability of the architecture.

The savings in energy when F_c is decreased from 1.0 to 0.8 is not considerable. Hence, a connection box with a flexibility of 1.0 is implemented. The pins of the logic block can connect to all the tracks in the corresponding routing channel.

The switch box is implemented with a flexibility of 3, using a disjoint switch box. This permits connection from each track to the corresponding tracks on the other three sides. This is similar to the switch box used in the XC4000 FPGAs.

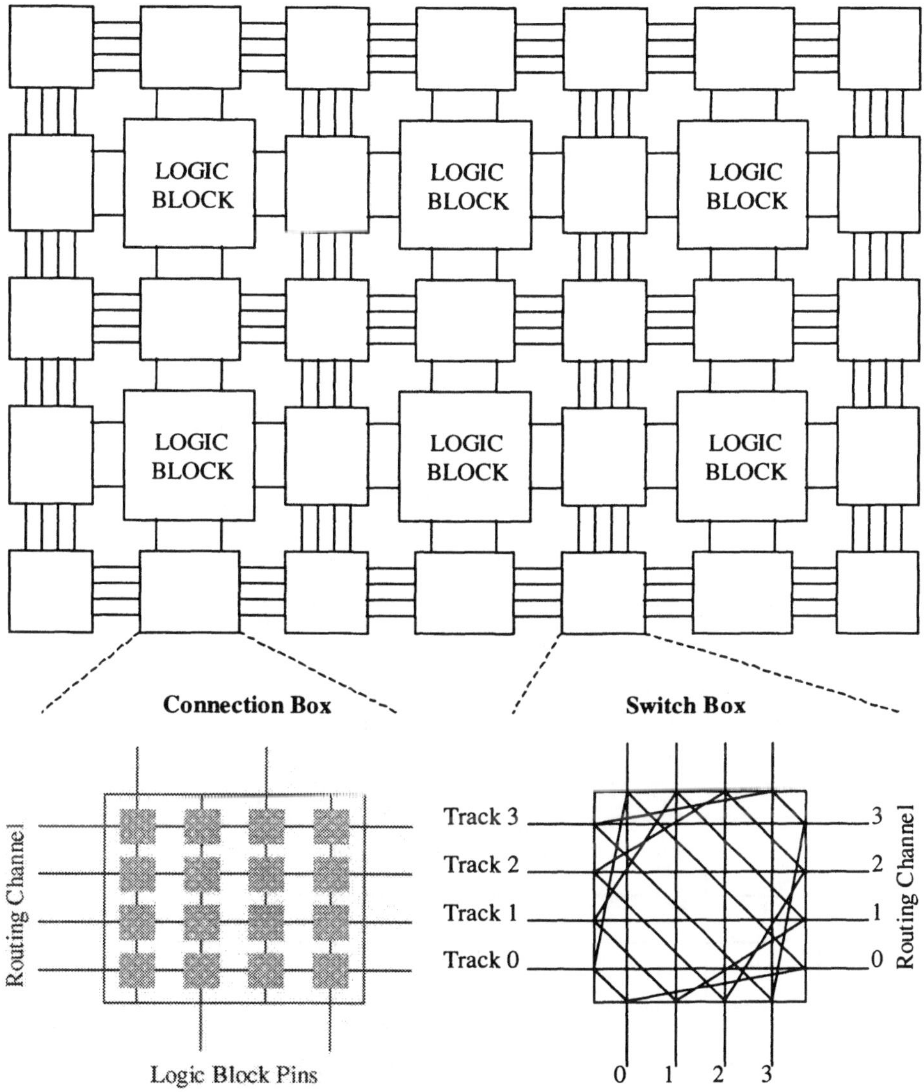

Figure 9. Level-1: Mesh Architecture

7.2 Nearest Neighbor Connection

For realizing local one-to-one connection between adjacent blocks, the Mesh structure is augmented with a level of nearest neighbor connections (NNC). This layer provides dedicated, low energy connections between adjacent logic blocks without having to go through the mesh routing structure. For example, Fig. 10 illustrates a neighborhood of eight, and twenty-four for the logic block *x*. Each of the output and input pins of the logic block can connect directly to the input and output pins, respectively, of the other blocks in the region.

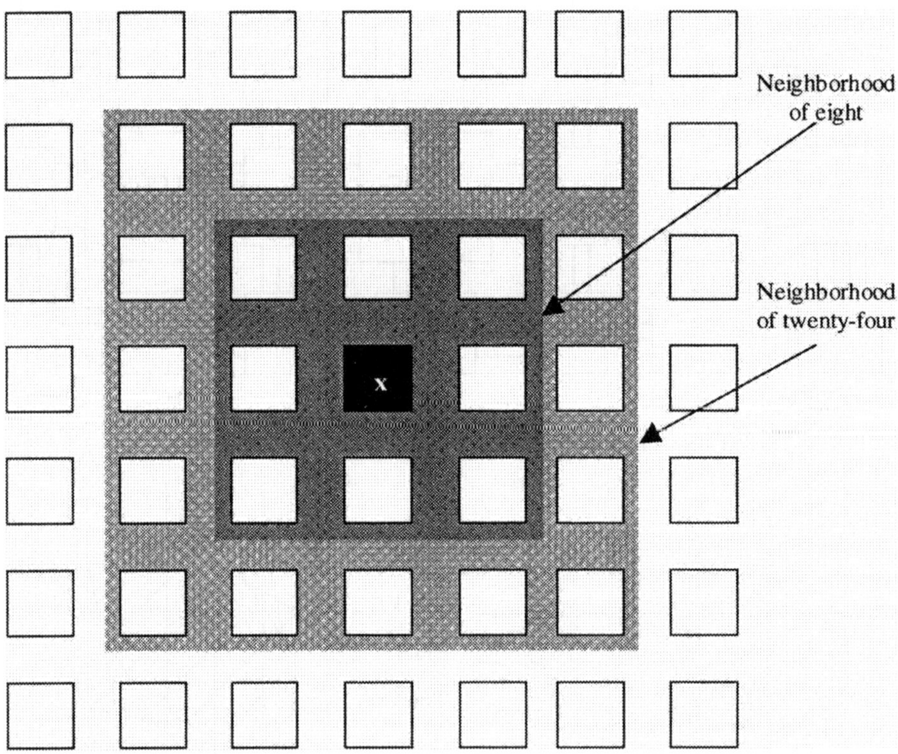

Figure 10. Level-0: Nearest Neighbor Connection

Each of these connections goes through only the two switches in the connection boxes at the output pin and the input pin at the two ends of the route, independent of the size of the neighborhood. The diffusion capacitance at each of the connection boxes, is proportional to the size of the neighborhood, and is the factor that limits the size of the region. For example, the connections illustrated in Fig. 10 will see a fan-out of eight and twenty-four for the regions of eight and twenty-four respectively. Hence, it can be seen that an arbitrarily large neighborhood can turn out to be more expensive than the mesh structure because of the higher fan-out.

The size of the neighborhood can be determined by comparing the cost of the mesh structure and the NNC as a function of the size of the neighborhood. For this comparison, the logic block used is the 5-input, 3-output structure that has been found to be optimal. This results in a capacitance of $10 \cdot C_{diff}$ for each of the single segments. The capacitive cost of each of the connection boxes is $W \cdot C_{diff}$, where W is the width of the channel. The cost of the connection boxes for the NNC is $(W+N_x) \cdot C_{diff}$, where N_x is the size of the neighborhood. The average energy-delay product for the different regions in terms of R_{on} and C_{diff} is given in Table 3.

Table 3. Average Energy-Delay Product of Connections in Mesh and NNC Structures

Neighborhood Size (N_x)	Mesh Structure	NNC
2	$1000 R_{on} C_{diff}^2$	$392 R_{on} C_{diff}^2$
4	$1000 R_{on} C_{diff}^2$	$648 R_{on} C_{diff}^2$
8	$1750 R_{on} C_{diff}^2$	$1352 R_{on} C_{diff}^2$
24	$4650 R_{on} C_{diff}^2$	$6728 R_{on} C_{diff}^2$

The aim is to maximize the size of the neighborhood as long as the Level-0 connections are cheaper than the Mesh structure. It can be seen that as the size is increased past eight to twenty-four, the NNC is more expensive than the Mesh. Therefore a neighborhood of eight is chosen.

7.3 Inverse Clustering

The energy and delay of a route in a Mesh structure can be computed in terms of the number of segments, n, and is given by,

$$Energy(E) \propto n \cdot C_{seg}$$

$$Delay(t) \propto \frac{n \cdot (n+1)}{2} R_{on} C_{seg}$$

In a Mesh structure, the number of segments in series increases linearly with the Manhattan distance, *d*, between the logic blocks to be connected. This means that the delay of the interconnect has a square dependency on the Manhattan distance, and the energy-delay product degrades as a cube function of the distance, as illustrated in Fig.11.

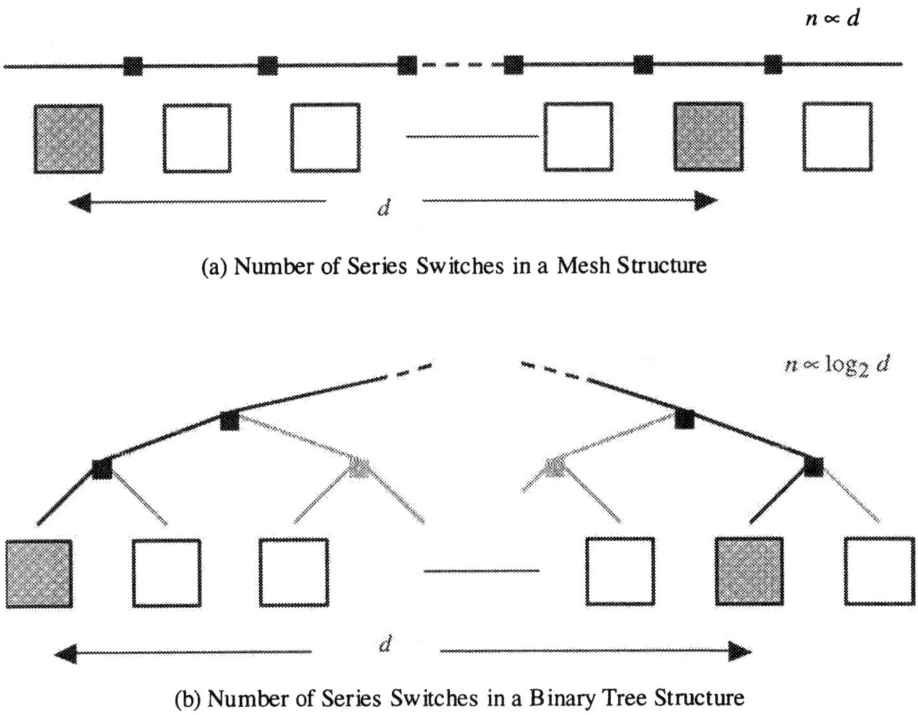

Figure 11. Mesh vs. Binary Structure

Different interconnect structures have been proposed to overcome this linear dependency of the number of switches to Manhattan distance. A hierarchical binary tree structure has been proposed in other works as a solution to this problem [Tsu99][Lai98]. An advantage of a binary tree connectivity is that the number of switches in series in a route connecting two logic blocks increases as a logarithmic function of the Manhattan distance. This is illustrated in Fig. 11.

Fig. 12 compares the Energy-Delay product of the Mesh structure and the Binary tree connectivity as a function of the Manhattan distance, for an array size of 16 x 16 logic blocks. For the Mesh architecture, a full connectivity in the connection box is assumed and the flexibility of three is used for the switch box. The parasitic contribution of the switches and the metal traces is based on a 0.25μm CMOS process.

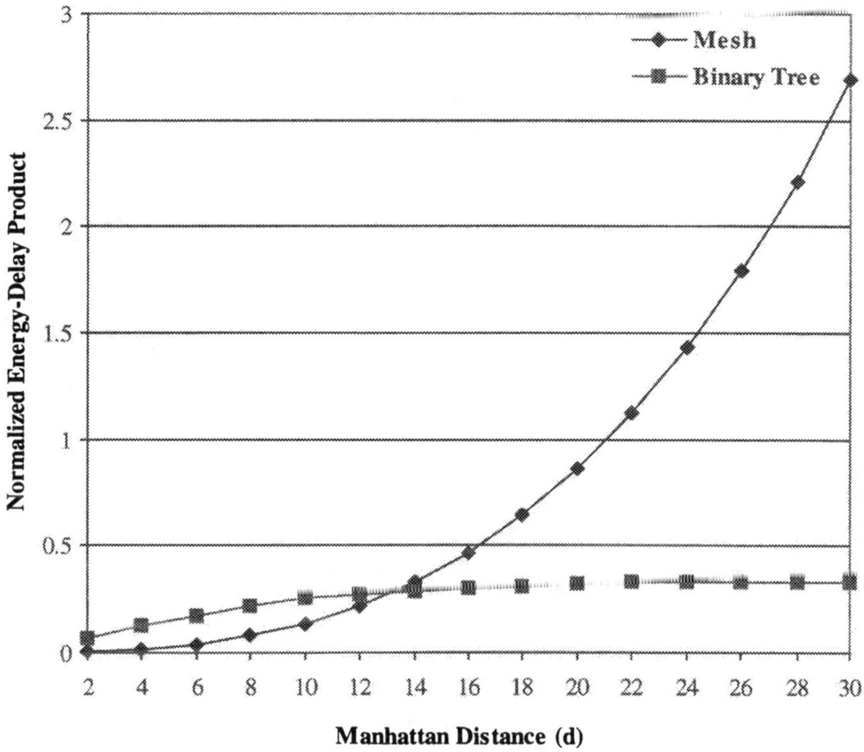

Figure 12. Energy-Delay Product as a Function of Manhattan Distance

For the Binary structure, the analytical model developed in [Lai97] is used. This model can be used to compute the flexibility of the switch boxes used in each node of the hierarchical tree. For a given *m*-ary tree structure, the number of switches and the flexibility of the switch box are functions of the required routability. The switch population is computed based on a Rent's coefficient of 0.67.

For the 16x16 array studied, for Manhattan distances up to ~13, the Mesh architecture can route connections that are cheaper than the Binary tree-like interconnect structure. This crossover point is a function of the size of the array and the specific process technology used to compute the cost of the interconnect.

The Mesh structure is more efficient for routing short wires, while the Binary tree-like interconnection structure is better for routing long wires. To exploit this characteristic, a combination of Mesh and a Tree structure is best suited to route all of the wires. The Mesh can be used to route the short wires, while the Tree structure can be used to route the long wires that span a significant fraction of the array.

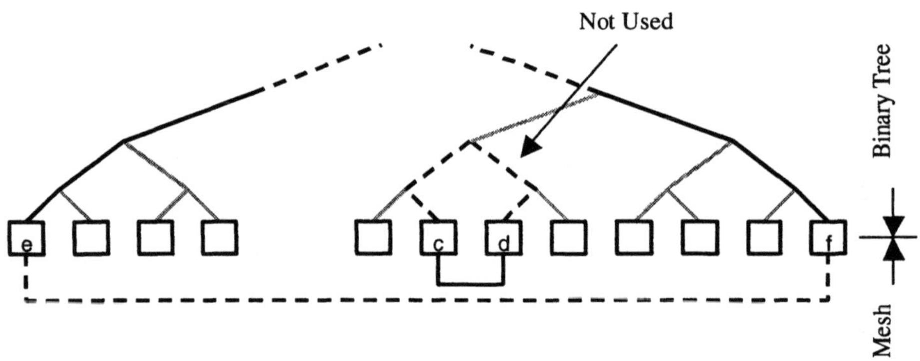

Figure 13. Connections in a Mesh + Binary Hybrid Architecture

Further refinement in the tree structure can be made by observing the clustering employed in a typical binary tree architecture. As can be seen in Fig. 13, the logic blocks close to each other are connected at the lowest levels of the tree, while a connection between logic blocks further apart has to traverse multiple levels of the tree structure. The routes that have to traverse fewer levels of the tree are cheaper in terms of the Energy-Delay product, since they encounter fewer switches. In a simple hybrid structure as shown in Fig. 13, the cheaper connections in the Binary tree will never be used since the Mesh is cheaper for realizing the short connections.

By modifying the clustering of the logic blocks, a better match can be obtained in the hybrid structure, as illustrated in Fig. 14. The clustering is such that the logic blocks further apart can be connected by going through fewer switches.

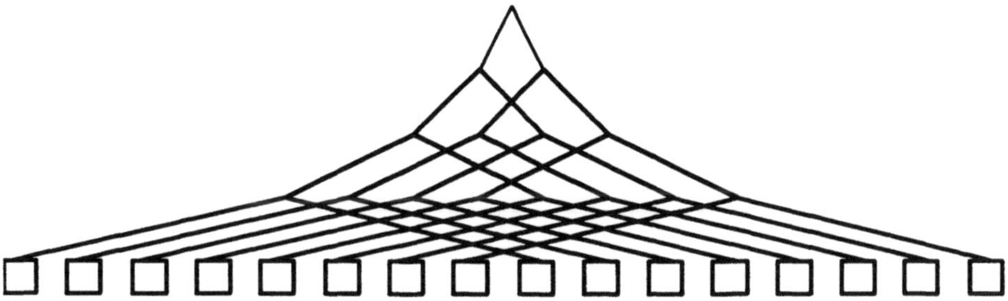

(a) Inverse Clustering

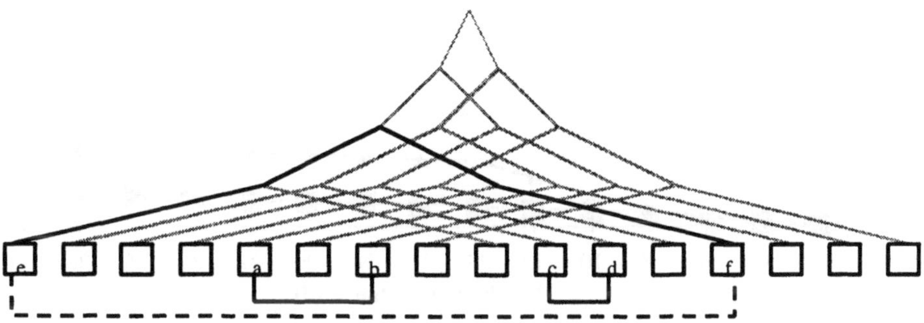

(b) Hybrid Architecture Using Inverse Clustering

Figure 14. Level-2 Inverse Clustering

This structure complements the Mesh structure in the hybrid architecture. The close connections can be routed in the Mesh structure, while the longer connections can be routed in the lower levels of the Inverse Clustered tree structure. This ensures that all of the cheap connections in the hybrid architecture are fully utilized. Fig. 15 shows the Energy-Delay product of connections routed in the hybrid architecture using inverse clustering structure. The Energy-Delay performance is contrasted with that of a hybrid architecture employing a traditional clustering. It can be seen that the inverse clustering mechanism attains a significant improvement over traditional binary clustering.

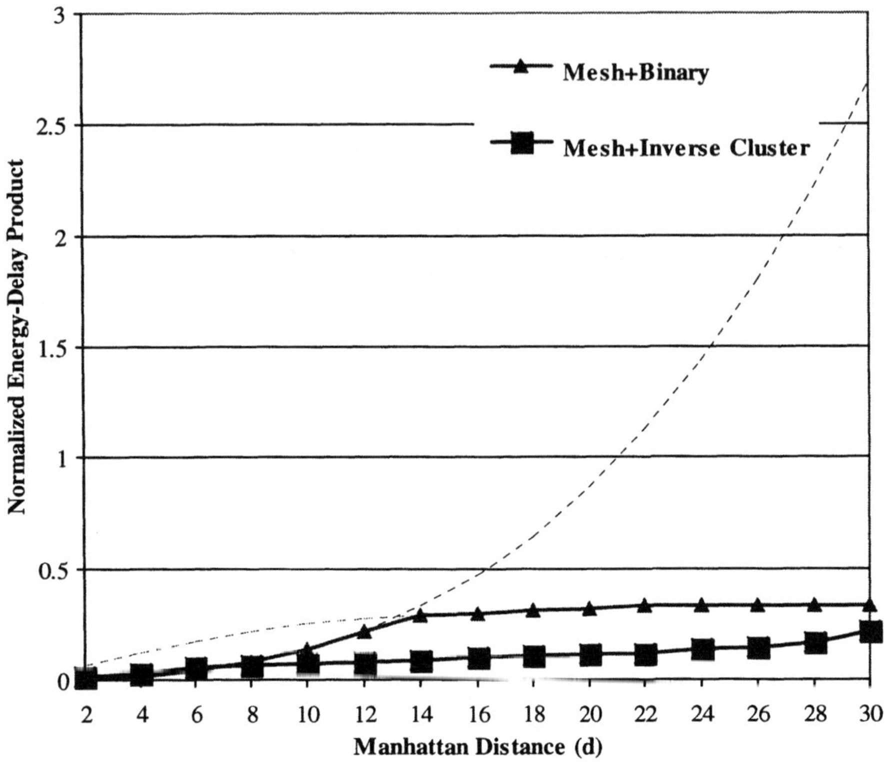

Figure 15. Energy-Delay Product of Hybrid Architecture

8 CONCLUSION

The architectural modifications have been mainly targeted at reducing the performance overhead of the interconnect. Optimization of the logic block helps to improve the interconnect utilization. The optimal logic block was found to be a 5-input, 3-output structure capable of implementing a 5-input random logic, or a 2-bit arithmetic operation.

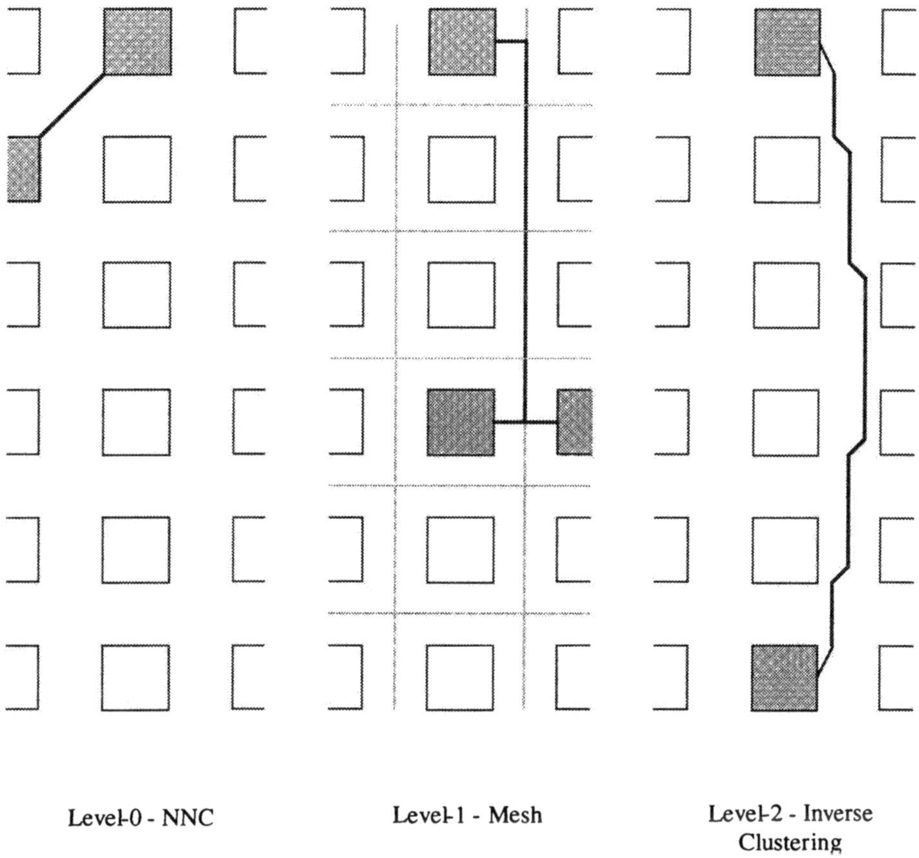

Figure 16. Interconnect Levels

The interconnect architecture is composed of three different structures, each of them targeted to a specific class of connection length. This is illustrated in Fig. 16.

- Level-0: Connects adjacent blocks using Nearest Neighbor connects.
- Level-1: Supports intermediate length wires on the Mesh structure.
- Level-2: Routes long nets spanning a significant fraction of the array using the inverse clustering structure.

For larger arrays it is quite possible that the three-level architecture discussed in this chapter will no longer be optimal. In such a scenario, further levels of hierarchy have to be considered.

Chapter 5

CIRCUIT TECHNIQUES

1 INTRODUCTION

Optimizations at the architectural level must be adequately supported at the circuit level. The interconnect is the dominant component of the total energy consumption, with the clock distribution coming in second. Since the routing switches dominate the energy and delay performance of the interconnect, the design of the switches is crucial. Low-swing signaling is another technique which can reduce the interconnect energy. Most of the low-swing techniques are targeted at busses and similar structures, where the capacitance of the interconnect is known. The applicability of these techniques in an FPGA environment has to be explored.

The energy contribution from the clock is significant. The distribution of flip-flops in an FPGA is sparse, and quite regular compared to an ASIC design. Energy reduction techniques can be used to exploit these characteristics.

2 RELATED WORK

Published material on the circuit and implementation issues related to FPGAs is quite limited. Most of the original work on FPGA design has been in the commercial sector, and as a result the information is proprietary. Some recent work has looked at the effect of circuit design on the performance of FPGAs.

Chow, et al. looked at the circuit design aspects of an SRAM-based FPGA [Chow99a][Chow99b]. Recognizing the capacitive load of the interconnect switches, work was done on restructuring the C box to reduce its effect on the routing channel. No major modifications were done on the interconnect. The threshold drop across the pass transistors in the routing fabric introduces the problem of leakage current in the gates following the switches. This problem was addressed by adjusting the switching voltage of the gates. This solution is technology dependent, and cannot be effectively carried on to future processes. Simulated data showed speed and area comparable to commercial structures.

Betz, et al. examined the transistor sizing and wire layout of FPGA interconnects from a speed performance perspective [Betz99a][Betz99b]. The gate-boosting method used in commercial architectures was adopted to address the threshold drop problem. This method of using a higher voltage on the gate of the transistors often requires additional circuitry. It was shown that the spacing of the routing wires has a significant impact on the speed performance of the routing. This can be attributed to the fringe capacitance of the wires that is becoming dominant in each successive process generation. The concept of electrically heterogeneous routing is introduced. This method advocates optimizing some of the routing for speed and the rest for density.

3 ENERGY-DELAY DESIGN SPACE

The connections in an FPGA are controlled by the connection box and switch box. In a typical SRAM programmable FPGA, the connections are realized using NMOS pass transistors that are controlled by SRAM cells connected to the gate terminal as shown in Fig. 1. By writing a logic one or a logic zero into the SRAM cells, the pass transistors can be switched ON or OFF to control the connections.

It was shown in Chapter 4 that the energy and delay of an interconnect route can be modeled in terms of the capacitance and resistance of the switches and the metal traces. Since the series resistance in the network is dominated by that of the pass transistors, one way to improve the speed is by making the transistors wider and reducing the resistance. The drawback with this approach is that the increase in transistor width is accompanied by an increase in the diffusion capacitance. It was shown earlier that the capacitive load in the interconnect is dominated by the diffusion capacitance of the switches accessing the tracks. The improvement in speed has to be paid for with increased interconnect capacitance, and hence increased energy.

Fig. 2 gives the energy and delay of a typical route in an FPGA as a function of the transistor width in a 0.25μm CMOS process. It is seen that the energy consumption of the route increases linearly with the transistor width. The improvement in the delay flattens out as the transistor width becomes very large. For the design of the low-energy FPGA, the sizing of the transistors is based on the energy-delay product. This ensures that the delay of the interconnect is not sacrificed to obtain the minimum energy.

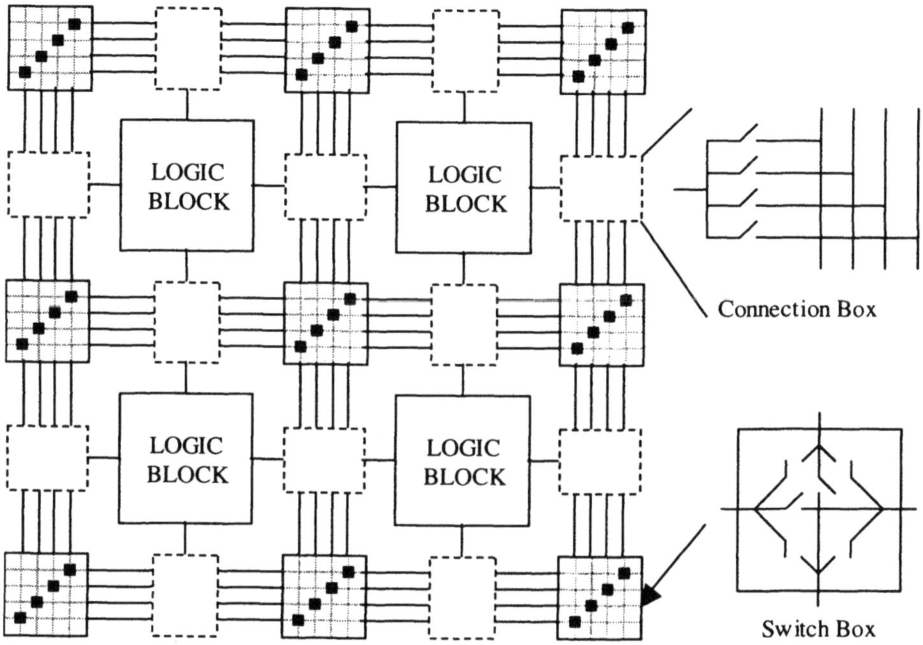

(a) Construction of Connection Box and Switch Box using Switches

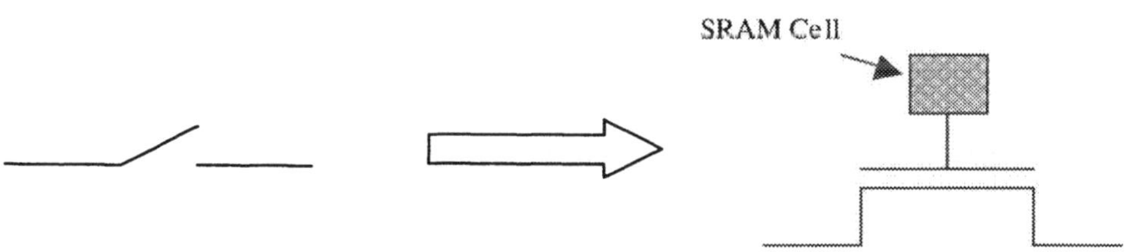

(b) Realization of Switches using NMOS Pass Transistors

Figure 1. Construction of Connection Box and Switch Box

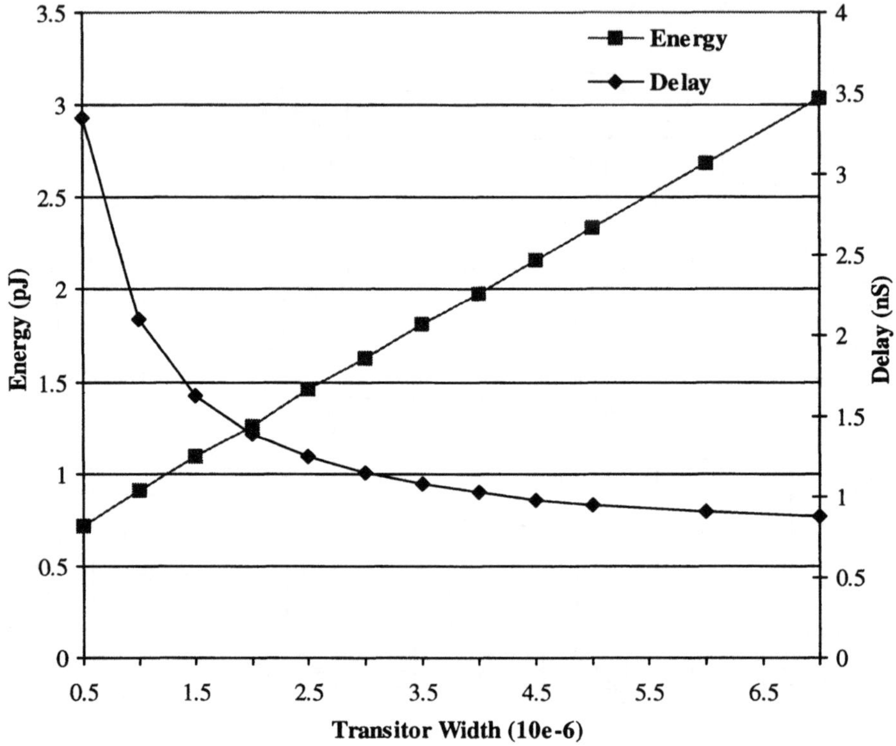

Figure 2. Energy and Delay of a Route in an FPGA as a Function of Transistor Width

4 LOW-SWING SIGNALING

The dynamic energy dissipated when driving a load is given by,

$$E_{dynamic} = C \cdot V_{dd} V_{swing}$$

Where,

C	is the capacitive load being charged
V_{dd}	is the supply voltage from where the current is drawn
V_{swing}	is the voltage swing on the load

By reducing the supply voltage and the voltage swing on the interconnect, a significant reduction in the total energy consumption can be achieved.

Considerable research has been done in the field of low-swing signaling, and a compilation of the different methods can be found in [Zhang00b]. These methods employ different circuit techniques and technology features to maintain signal swings in the sub-one volt region. The techniques have been targeted at the general ASIC environment. The viability of these techniques in an FPGA environment has to be evaluated.

4.1 Existing Low Swing Techniques and the FPGA

The basic principle of some of the existing low swing techniques can be broadly classified based on their voltage sensing schemes.

4.1.1 Low Threshold Devices

It is easier to produce a low-swing signal, than to detect the signal. Hence, it is always the limitations of the receiver that constrains the minimum voltage swing possible. If a complementary static circuit is to be used as the receiver, the threshold voltage of the devices is the limiting factor. One way of improving the sensitivity is by using low-threshold devices in the receiver circuit [Nakagome93].

The problem with this method is the use of special low-threshold devices used in conjunction with standard threshold devices. This requires additional steps during device fabrication tuned just for the FPGA.

4.1.2 Differential Interconnect

Apart from the sensitivity of the receiver circuit, the required noise margin places a lower bound on the minimum voltage swing. This is required to avoid erroneous switching due to noise injected into the system. The immunity of the receiver circuit to common-mode noise can be improved by using a differential sensing scheme. Since the common mode noise affects both the signals equally, they can be cancelled out. This method has been proposed by [Liu94] to achieve low voltage swings.

The problem with this approach is that a single signal requires two wires. This can add considerable area overhead in a system like an FPGA, where the dominant area component is the interconnect. This would further reduce the logic density of the FPGA, making it a poor choice for most applications.

4.1.3 Timing Signals

Another method to detect low signal swings is by using a sense amplifier. The drawback of traditional sense amplifiers is the constant tail current of the circuit. One way of reducing this constant current is by enabling the sense amplifier only when required. This can be done if the load being driven is known exactly. In such an environment, the time taken to drive the load is known, and the sense amplifier can be activated using a timing signal. This method has been used to control the swing and the receiver circuit [Colshan94].

The main drawback to this method is the assumption that the load is known. In an FPGA, the capacitance being driven is dependent upon the route taken to realize a connection. Different nets will have different lengths, and hence different capacitances. This rules out the use of a single timing signal to activate the receivers on the different nets.

4.1.4 Charge Recycling

Charge recycling or charge sharing methods of achieving low swing and quadratic energy saving is a novel method [Hiraki95] [Yamauchi95]. As the name implies, the methods employ charge sharing between the multiple data-lines of a bus to reduce the voltage swing by a factor of n, where n is the number of data-lines in the bus. The energy dissipation is always constant, independent of the signal activity on the bus. One of the main requirements for this method is that the capacitive load on all the data-lines in the bus must be closely matched.

The interconnect in an FPGA does not normally use bus-based data transfer. Even if bus-based transfer is possible, for special architectures using coarse granularity logic blocks, like datapath units, it will be difficult to satisfy the condition of capacitive matching.

5 LOW-SWING CIRCUIT

The low-swing signaling circuit proposed for use in the FPGA is shown in Fig. 3. The driver translates the voltage from V_{DDH} to V_{DDL}. The interconnect is made up of the NMOS pass-transistor switches with the control voltage, V_{DDC}, on the gate. The receiver translates the low voltage signal back to V_{DDH}.

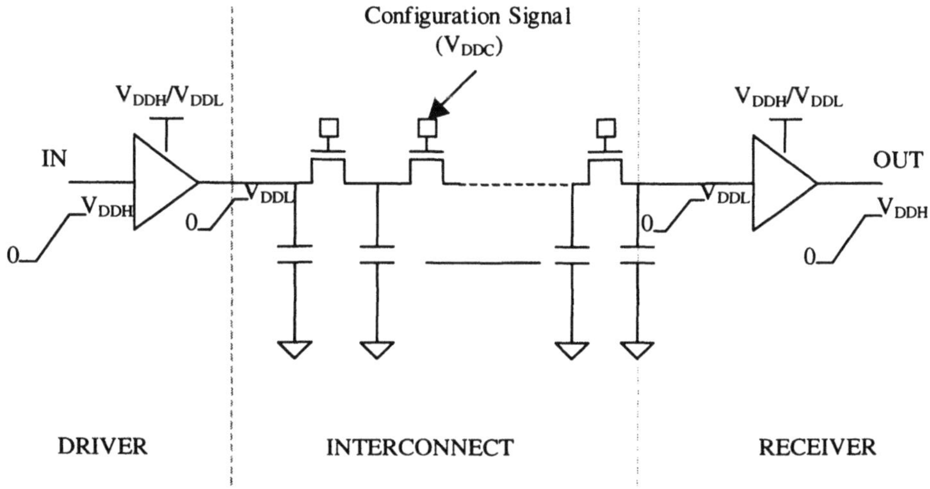

Figure 3. Low-Swing Circuit

5.1 Driver

The main purpose of the driver circuit is to drive the capacitive load at the reduced swing of 0-V_{DDL}. Fig. 4 shows the two basic driver configurations that can be employed.

In the first configuration, the output voltage swing will be from 0 to V_{DDL}, which is the required high voltage. If the voltages are such that $V_{DDH} \geq V_{DDL} + V_{Tn}$, where V_{Tn} is the threshold voltage of the NMOS device with body effect, then the second configuration can also support a voltage swing between 0 and V_{DDL}.

If the voltage relationship between V_{DDH} and V_{DDL} can be ensured, then the second configuration is preferred. For the same gate voltage, an NMOS transistor has a higher current driver per unit width than a PMOS transistor. Hence, the second configuration will have a smaller area than the first configuration for the same delay specification. The overhead of the extra inverter is negligible compared to the size of the output devices.

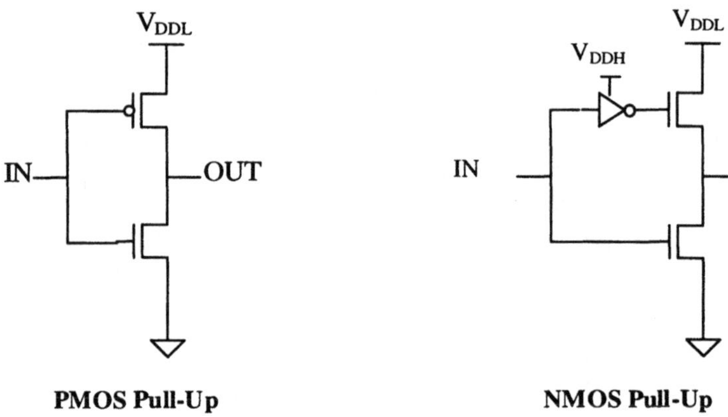

Figure 4. Driver Configurations

5.2 Receiver

In a low-swing circuit, the receiver is the critical component. The circuit has to be able to sense the low voltage at the input. In the absence of low-threshold devices, timing signals, and differential signaling, one method is to use a pseudo-differential sensing scheme. The input low-voltage signal can be compared to a reference signal using a differential amplifier configuration. The drawback of this method is the constant tail current. Since all of the logic blocks have five inputs, the static current from receiver blocks in the entire array will be unacceptable.

The receiver circuit used in the proposed scheme is shown in Fig 5. This circuit is completely static, and does not require timing signals. When the input to the receiver is low, Node 1 is charged to V_{DDH}. During a low to high transition, MN1 is switched ON, pulling Node 2 low. This triggers the discharge of Node 1 through MN2, which acts as a cascode amplifier. The transition at Node 1 activates MP2 that pulls the output node high. During a high to low transition, MN3 pulls the output low enough to trigger MP1. During both transitions, MP1 and MP2 form a differential pair to obtain faster transitions between the stable states. This helps to reduce the short-circuit currents in the receiver circuit.

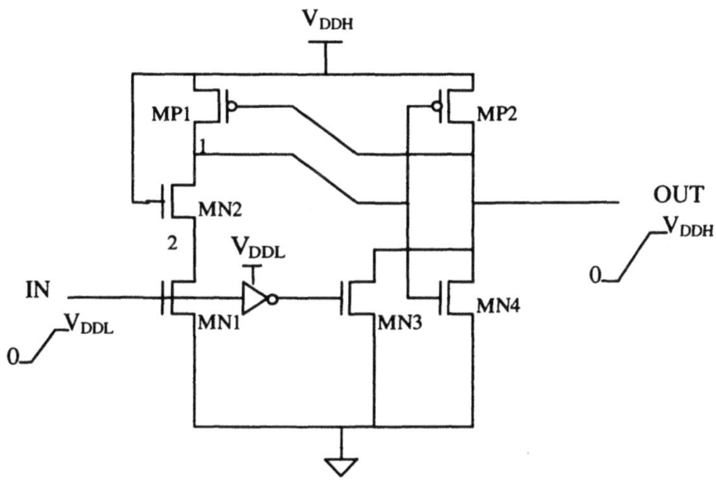

Figure 5. Low-Swing Receiver

5.3 Configuration Voltage

V_{DDC} represents the control voltage applied to the gate of the NMOS transistors used as routing switches in the interconnect network. The power supply voltage of the SRAM cells used to store the configuration determines this voltage. V_{DDC} has the same constraints as imposed on V_{DDH}, to avoid the threshold voltage drop across the routing switch.

The R_{on} of the switches is dependent on the gate voltage. By keeping the gate voltage high, R_{on}, and hence the interconnect delay, can be reduced. Since the control voltage is steady during normal execution, there is no penalty in terms of energy due to higher V_{DDC}. The delay of a typical interconnect route as a function of the control voltage is given in Fig. 6. For an interconnect swing of 0-0.8V, the delay can be improved by almost a factor of two by boosting the control voltage from 1.5V to 2.5V.

The disadvantage of having an independent V_{DDC} is the requirement of another power supply. For this work V_{DDC} is kept the same as V_{DDH}. The advantages of having a higher V_{DDC} can be exploited if required.

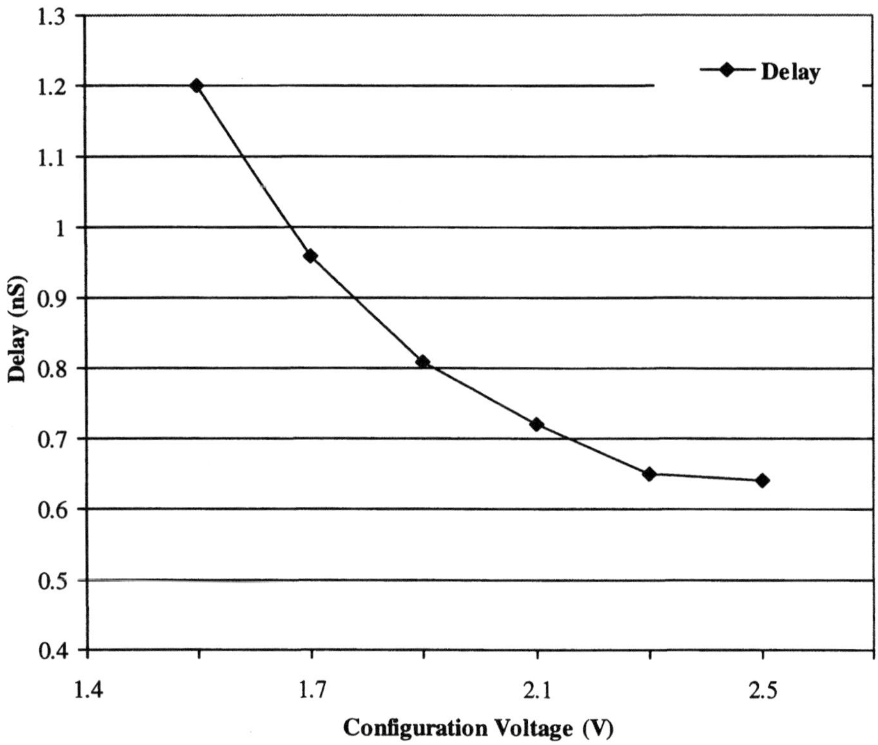

Figure 6. Interconnect Delay as a Function of Configuration Voltage with V_{DDL}=0.8V

5.4 Performance

The circuit shown in Fig. 3 is simulated with a typical interconnect route. For the given process technology, V_{DDH} and V_{DDL} were chosen to be 1.5V and 0.8V, respectively. The circuit is compared against a full-swing circuit running at 1.5V. The comparison shown in Table 1 includes the overhead of the driver and receiver circuits. There is a significant improvement in energy, and the energy-delay product is smaller by almost a factor of two.

Table 1. Comparison Between Full-Swing and Low-Swing Circuits

	Full-Swing Circuit V_{DD} = 1.5V	Low-Swing Circuit V_{DDH}=V_{DDC}=1.5V, V_{DDL}=0.8V
Delay (nS)	1.9	2.3
Energy (pJ)	72.3	31.4
ED Product	137	72

The overhead incurred in this method is an additional supply rail for the low voltage. DC-DC converters are quite common in most of the current digital systems to supply multiple voltages. Hence, the additional supply rail will not pose a big problem. In the extreme case of having to generate a separate supply for just the FPGA, it is possible with minimal overhead [Burd00].

6 CLOCK DISTRIBUTION

The clock distribution network is the next major component of energy consumption after the interconnect. The distribution energy can be reduced by evaluating the different components involved, and taking advantage of the characteristics of the FPGA environment.

6.1 Components of Clock Distribution Energy

The capacitance switched during the distribution of the clock can be divided into three major components: global distribution network, local distribution network, and the flip-flop load.

6.1.1 Global Distribution Network

The global clock network distributes the clock to the periphery of the logic blocks. For a given technology, C_{global} is a function of the area of the chip, and for a given array, C_{global} is dependent on the size of the logic block. For most applications, the global energy is fixed. For some applications that use a small fraction of the available flip-flops, it is possible to gate the clock to the branches of the distribution tree to reduce the energy.

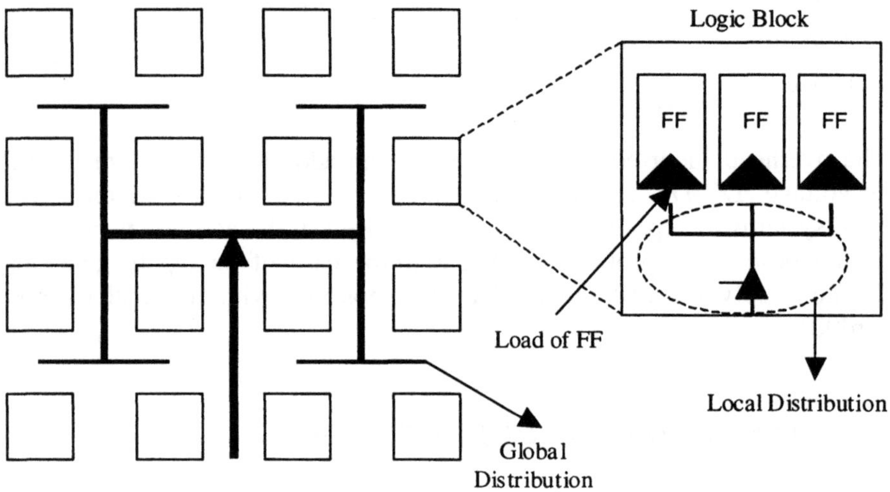

Figure 7. Components of Clock Distribution

6.1.2 Local Distribution Network

The local clock network distributes the clock from the global network to the flip-flops in the network. For a given technology, C_{local} is a function of the physical dimension of the logic block. If the flip-flops in a logic block are not used, the local distribution can be gated.

6.1.3 Flip-Flop Load

C_{ff} is the capacitive load added to the local clock distribution network by the flip-flops. This capacitance is dependent on the complexity of the flip-flops.

6.2 Clock Distribution in an FPGA

Each of the logic blocks has flip-flops associated with it. This necessitates distributing the clock over the entire array. Since the flip-flops are distributed sparsely over the entire array, the contribution of the distribution network dominates the total clock energy.

The contribution of the different components can be evaluated by designing a simple clock distribution network. The clock distribution is designed for a 16 x 16 logic block array in a 0.25μm CMOS technology. The area of each logic block is determined from actual layout information to be 250μm x 250μm. The global clock distribution is done as a balanced H-tree, with distributed drivers. The local clock distribution is approximated with a metal trace the length of the logic block side. The flip-flop used is a D-flip-flop from the standard cell library. The capacitive contributions of the components are given in Table 2.

Table 2. Contribution of the Clock Components

Clock Component	Equivalent Switched Capacitance
Global Distribution. Includes the cost of the distributed drivers.	10.3 pF
Local Distribution per logic block. Includes the cost of the driver.	42 fF
D Flip-Flop per logic block.	18.3 fF

It can be seen that for a 16 x 16 array of logic blocks with all the flip-flops being utilized, the capacitance from the flip-flops is only 20% of the total. The rest of the energy contribution is from the clock distribution network. As the utilization of the flip-flops reduces, the percentage contribution of the distribution network increases.

6.3 Clock Energy Reduction Methods

Reducing the number of flip-flops would appear to be a reasonable step in the reduction of clock energy. However, previous research in the depopulation of flip-flops [Rose90a] has shown this method to be counterproductive, because of the overhead incurred when data needs to be registered.

Based on the characteristics of the FPGA environment, and the preliminary capacitance data, it is clear that the distribution network has to be targeted for maximum energy reduction. Since the area of the logic blocks determines the distribution network, the absolute capacitance of the network cannot be reduced. In view of the fixed capacitance, the energy can be reduced only by reducing the voltage swing and the transition activity on the distribution network.

6.4 Low-Swing Signaling

The low-swing signaling technique presented in section 5 can be used to distribute the clock signal on the global distribution network. Since the logic blocks operate at the high voltage, the swing on the clock net is restored at the boundary of the logic block. This method will reduce the energy of the global distribution by approximately a factor of two.

6.5 Double-Edge-Triggering

Conventional single–edge-triggered flip-flops are used traditionally in designs. The data are sampled at only one clock edge, and the output data change only once per clock cycle. The implementation of a simple single-edge triggered flip-flop using transmission gates is shown in Fig. 8(a).

In a double-edge-triggered flip-flop, the input data are triggered at both edges of the clock, and the output changes twice in each clock cycle. This method allows halving the clock frequency while maintaining the same data-throughput. Since the energy consumed by the clock distribution network is directly proportional to the clock frequency, this flip-flop will result in halving the clock distribution energy. The implementation of the double-edge-triggered flip-flop is shown in Fig. 8(b) [Llopis96][Dally98].

The clock signal is loaded by eight and twelve transistor gates in the single-edge-triggered and double-edge-triggered implementations, respectively. Even though the clock load is higher, since the clock activity is halved, the double-edge-triggered flip-flop actually consumes less energy in the clock.

The energy dissipated due to the data is higher in the double-edge-triggered flip-flop due to the increased complexity. The total energy consumed in the double-edge-triggered flip-flop varies from 45% to almost the same as that of a single–edge-triggered flip-flop, as the data activity is varied from zero to one [Llopis96].

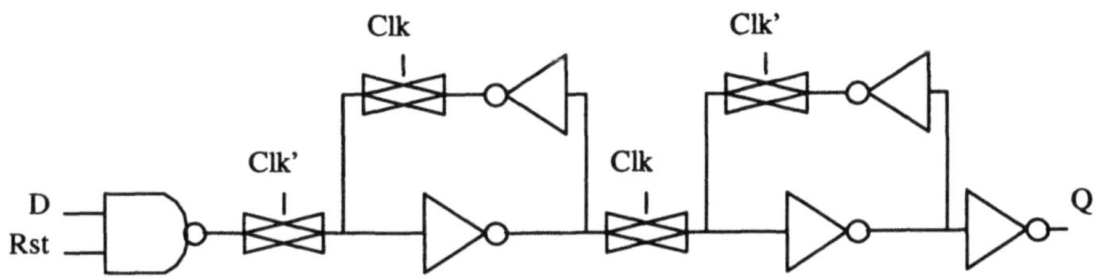

(a) Single-Edge-Triggered FF

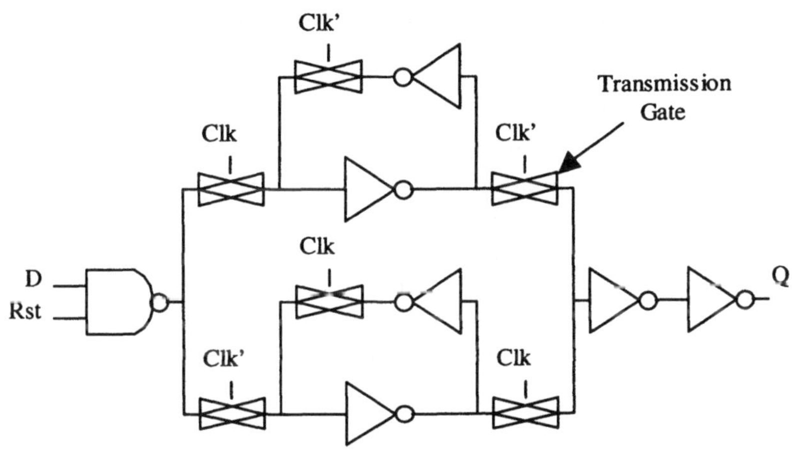

(b) Double-Edge-Triggered FF

Figure 8. Flip-Flop Styles

The single-edge-triggered implementation requires twenty-six transistors, while the double-edge-triggered implementation requires thirty-two transistors. This includes the two inverters used to buffer the clock signal and generate the complement. The increased area due to the increased number of transistors is one of the reasons why the double-edge-triggered flip-flop is not heavily used in traditional designs. In an FPGA, the area is dominated by the interconnect resources. The contribution of the flip-flops to the total area is approximately one percent, and negligible. Hence, the increased area of the double-edge-triggered flip-flop will not affect the total area.

The single-edge-triggered flip-flop is insensitive to the duty cycle of the clock to a large degree. This is not the case with the double-edge-triggered version, where the duty cycle has a direct effect on the data-throughput. Hence, the duty-cycle has to be maintained at 50%. The set-up time and the hold time of the double-edge-triggered flip-flop is comparable to the conventional flip-flop designs.

7 CONCLUSION

The architectural modifications described in Chapter 4 can be supported with circuit level optimizations to reduce the total energy consumption of the FPGA architecture. The modifications were aimed at the routing structure and the clock distribution network. Low swing signaling suitable for deployment in an FPGA environment is demonstrated with ~2x improvement in energy. The clock distribution structure incorporates the same low-swing technique to reduce the clock energy. A double-edge-triggered clocking strategy is employed to halve the transition activity on the distribution network with corresponding energy gains.

The proposed circuit techniques do not require special processing steps, and can be implemented using a typical digital process technology.

Chapter 6

CONFIGURATION ENERGY

1 INTRODUCTION

The realization of an application in the FPGA has two steps, configuration and execution. During the configuration step, the memory cells that control the behavior of the logic and routing resources are loaded with the programming bits required to implement the required application. In the execution step, data are processed by the function implemented in the FPGA. The impact of the configuration step on the total energy and speed is dependent on how frequently the FPGA is programmed. When the FPGA is used as a performance accelerator, the array is reconfigured often enough for the configuration step to be considered as a significant component.

This chapter will look into the effect of the configuration technique on the energy and speed. The shift register technique and the random access technique represent two contrasting methods of programming the FPGA. The characteristics of the methods will be explored to evaluate their impact on the usage model of the FPGA. Based on results from a preliminary implementation of the random access technique, different modifications are explored to improve the energy performance of the configuration step.

2 CONFIGURATION COST

Each programmable component in an FPGA is controlled using data stored in SRAM cells. These memory cells, in turn, have to be loaded with valid data before the FPGA can be used. If the FPGA is to be used as a performance accelerator where rapid re-configuration is required, then the overhead associated with the configuration step has to be considered. The FPGA can be used as an accelerator unit only if the total cost of using the FPGA is less than a software implementation:

$$Cost_{HW} + Cost_{Config} < Cost_{SW}$$

Where,

$Cost_{HW}$	Cost of implementing function in hardware (FPGA)
$Cost_{Config}$	Cost of programming the FPGA
$Cost_{SW}$	Cost of implementing the function in software (GP)

In the above equation, $Cost$ is used to represent the delay or the energy consumed when the function is implemented.

As an example, consider the configuration overhead for the popular XC4000 architecture [Xilinx2]. For the XC4000XL series, the number of bits needed to program the FPGA ranges from 151,910 for the XC4005XL with a logic capacity of 466 4-input LUTs to 1,924,940 bits for the XC4085 with a logic capacity of 7,448 4-input LUTs. Even in the fast parallel programming mode, this overhead translates to approximately 19,000-240,000 write cycles. This is exacerbated by the fact that the write speed for programming the FPGA is slower than the system operation speed by a factor of approximately ten.

The high overhead of programming the FPGA limits the computation functions that can be efficiently implemented in the FPGA. Only those functions that have to process long streams of data can be considered as viable candidates for implementation in the FPGA.

3 CONFIGURATION TECHNIQUES

A better understanding of the cost of programming the FPGA can be obtained by comparing the two basic methods of loading the configuration: the shift register and the random access memory technique. Almost all commercial FPGA architectures use the shift register technique. This can be attributed to the fact that early FPGAs were not operated in an environment where the configuration overhead in terms of speed and energy was a major factor. The random access programming method offers a reduction on the speed and energy penalty. To get a complete picture, these methods have to be evaluated in terms of the area, pin-count, delay, and energy overhead.

3.1 Shift Register Technique

The shift register method of programming is the simplest. The configuration is fed in as a single bit-stream. The configuration storage cells are connected as a long chain, and the bit-stream is shifted into the array one bit at a time. This method is illustrated in Fig. 1. When the entire bit-stream is loaded into the array, each configuration cell will contain the correct data to program the switch to which it is connected. Hence, independent of the utilization of the array, the entire bit-stream has to be loaded in each time.

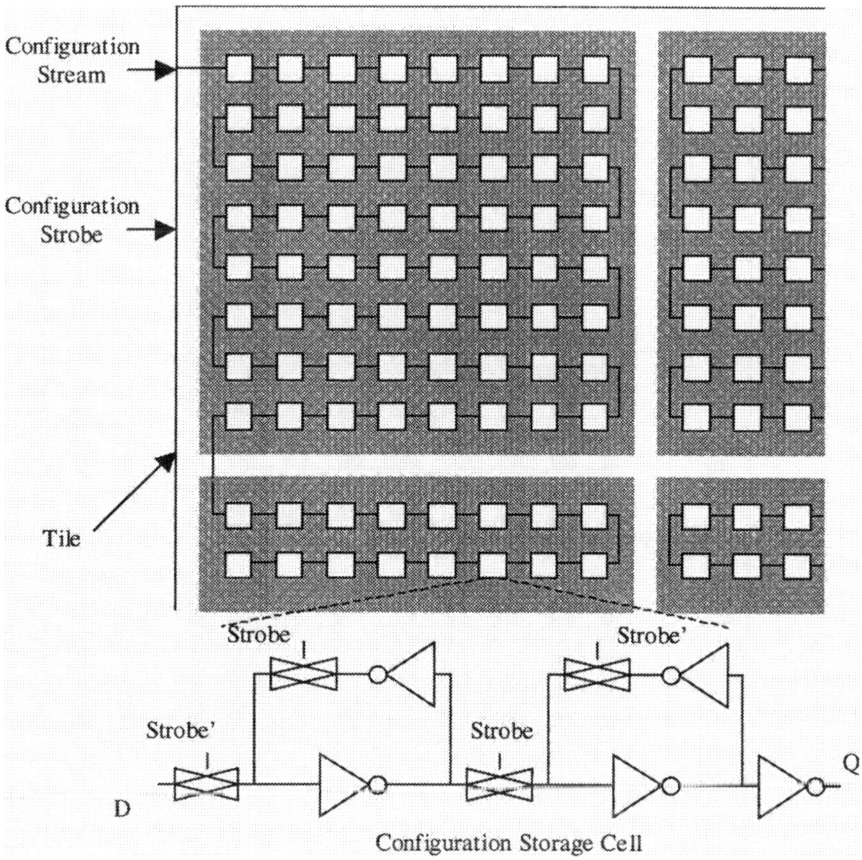

Figure 1. Shift Register Method of Programming

This method also has an impact on the cell that can be used to store the configuration. Each memory cell is used to not only store the configuration, but also to drive the next memory cell in the chain. This requires the use of at least a flip-flop as the configuration storage cell. The configuration storage cell is shown in Fig. 1. Almost all of the commercial architectures use this method to program the FPGA.

3.2 Random Access Programming

The random access technique is quite different from the shift register technique. This method is similar to writing data into a random access memory (RAM), and is illustrated in Fig. 2.

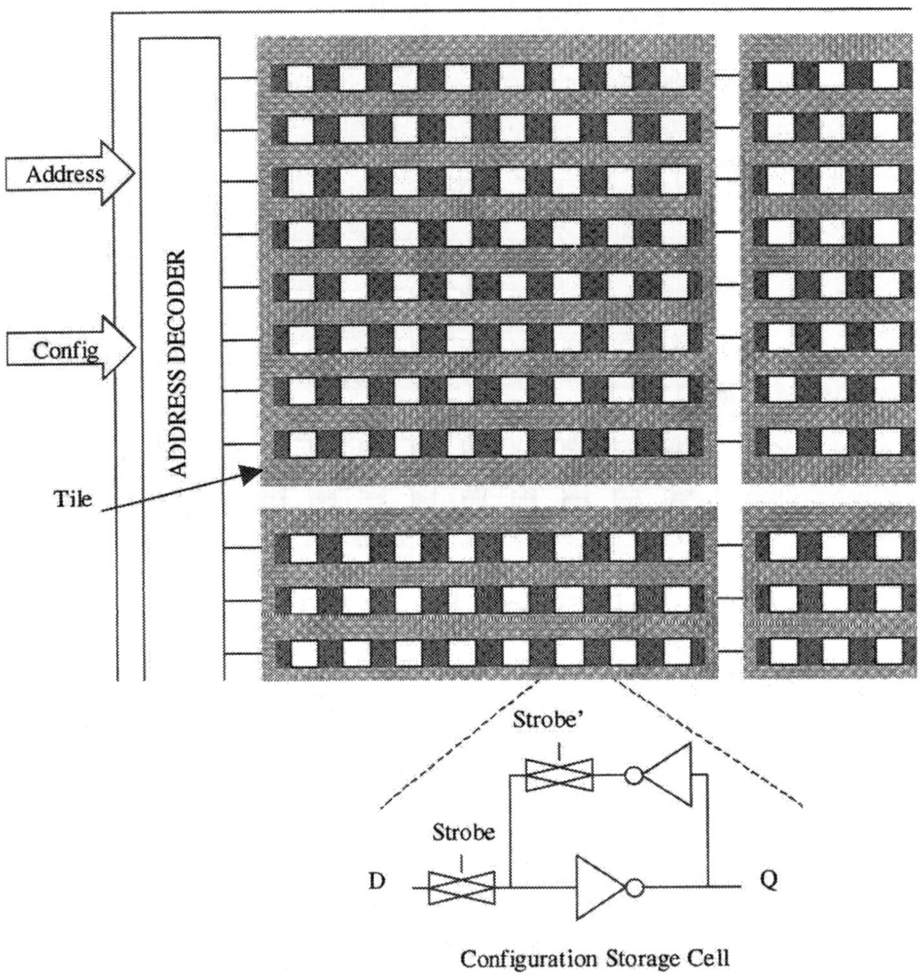

Figure 2. Random Access Programming

The configuration bits are grouped as words and can be selectively programmed. This allows programming of only the resources that are being utilized. However selective addressing requires additional circuitry for decoding, and necessitates the routing of configuration data and address busses through the entire array.

This method allows the use of a very simple cell for storing the configuration information. The functionality required of the cell is to hold the data to program the switch it is controlling. This can be achieved by using a simple latch as the configuration storage device. This is considerably simpler than the flip-flop used in the shift-register method. The programming method and the configuration storage cell are shown in Fig. 2.

4 SHIFT REGISTER VERSUS RANDOM ACCESS

The parameters that can be used to compare and contrast the two methods are pin count, partial programming, area, delay overhead, and energy overhead.

4.1 Pin Count

In the shift register method of loading configuration, only two pins are required, independent of the size of the array. One pin is for the configuration data, and the other pin is for moving the data through the register chain.

In the random access method of programming, the number of pins required for configuration is dependent on the size of the array. The number of pins for loading the configuration data is fixed and determined by the grouping of the bits. The number of pins for addressing the memory space is dependent on the number of configuration words in the array.

In a stand-alone FPGA, the number of pins dedicated to configuration loading is important since it has a large impact on the packaging. For an embedded FPGA, the number of pins dedicated for the configuration is not critical.

4.2 Selective Programmability

Selective programmability of the array is desirable for a number of reasons,

- The function being implemented utilizes only a fraction of the total array.
- All the logic and interconnect resources in a tile need not be programmed each time.
- Successive tasks are similar enough that only a fraction of the logic and interconnect needs to be reprogrammed.

The shift register method of programming is an all-or-nothing approach. The entire array has to be reprogrammed each time, independent of the size of the application. As the name indicates, the random access programming method facilitates programming random configuration words using configuration addresses, making selective programmability feasible.

4.3 Area

The main component of the total area of an FPGA is the configuration. The number of configuration bits is independent of the programming technique, but the method of programming can affect the total area.

The shift register programming technique requires the use of a master-slave flip-flop to store the configuration. The routing overhead is minimal since each flip-flop has to communicate only to the next flip-flop in the chain.

In the random access programming technique only the storage of the configuration is needed. A simple latch can be used for this purpose. Compared to a master-slave configuration, the area complexity of the latch is considerably lower. However, this programming technique requires an address decoder to select the configuration words, and the configuration data bus has to be routed through the entire array.

4.4 Delay and Energy Overhead

It is envisioned that using the FPGA as an accelerator can boost the performance of the system. In such an environment, the FPGA will be reprogrammed quite often. As the frequency of switching tasks increases, the overhead of the configuration step cannot be ignored.

The delay overhead of the shift register method is always fixed, and is dependent only on the size of the array, since the entire array has to be programmed each time. The energy overhead is dependent on the number of times each register is programmed. The energy overhead is seen to be proportional to N^2, where N is the configuration size of the array. Due to the serial propagation of configuration bits, N bits propagate through the first register in the chain, $(N-1)$ bits propagate through the second register, and so on.

To the first order, delay overhead of the random access programming technique is proportional to the number of configuration words that have to be loaded into the array. The energy overhead is also proportional to the number of words that have to be loaded into the configuration space.

4.5 Comparison of Techniques

Table 1 compares the two programming techniques. N is the configuration size of the array, W is the word size in the random access programming technique, and T is the configuration size of each task.

The optimal method of programming is dependent on the environment in which the FPGA is going to be employed. A stand-alone FPGA used as glue logic where rapid reprogramming is not an issue will benefit from the shift register method. An embedded FPGA used as a performance accelerator, requiring rapid and frequent reprogrammability, is better programmed using the random access technique.

Table 1. Comparison of Programming Techniques

Overhead	Shift Register Programming	Random Access Programming
Pin Count	2	$\mathrm{Log}_2(N/W) + W$
Area	$\propto N$	$\propto N$
Partial Programmability	No	Yes
Delay	$\propto N$	$\propto T$
Energy	$\propto N^2$	$\propto T$

5 CONFIGURATION ENERGY COMPONENTS

From the viewpoint of partial and selective programmability, the random access technique is the desirable choice. To get a better understanding of the costs involved, a simple implementation of the array is made. The configuration circuitry for an 8 x 8 array is designed as shown in Fig. 3. The configuration is grouped as 8-bit words and can be selectively programmed at this granularity. A hierarchical decoding structure is used to minimize the select lines traversing the array. In the first level of decoding, the row and column are selected to identify the tile being programmed. The location of the word is decoded inside the tile.

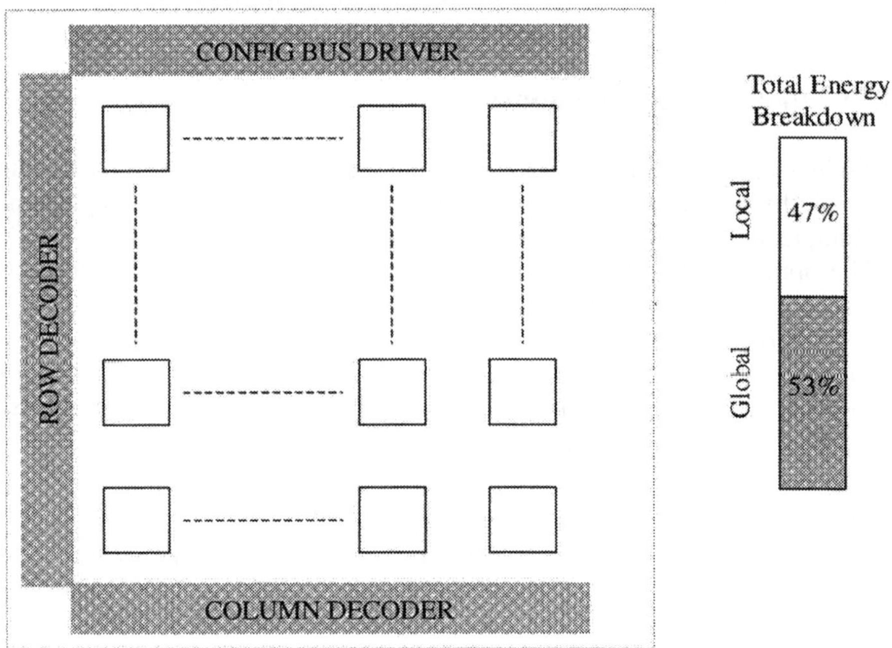

Figure 3. Configuration Energy Breakdown

By using the physical implementation, accurate performance data can be obtained. This method accurately accounts for physical parasitics. The parasitics include the capacitance of the metal wires and the capacitive load of the configuration storage cells.

The energy dissipated for each write can be divided into three components: global, local, and storage energy. The global energy takes into account the address decoders and the configuration bus drivers. The global energy includes not only the energy dissipated in the logic circuitry, but also the energy required to drive the control and data signals for the configuration across the entire array. This component is a function of the physical size of the array. Hence, it is dependent on the number of tiles in the array and the physical size of each tile.

The local energy is the energy dissipated in the tiles. This includes the energy dissipated for local decoding of the address and distributing the configuration data in the tile. The energy dissipation in the tile is dominated by the energy required to distribute the configuration data. The main components of this load are the interconnect capacitance and the capacitive load of the memory cells. Hence, this energy is a function of the number of configuration words per tile and the physical size of the tile.

The storage energy is the energy dissipated in the actual storage cells during each write. This is a function of the complexity of the memory cells, and the load presented by the switch that is being programmed.

The breakdown of the energy is shown in Fig. 3. The total energy is evenly distributed between the global and local components. The global energy accounts for ~53% and the local energy accounts for ~47% of the total energy. The interesting fact is that the storage energy is negligible and can be ignored. It accounts for barely 0.5% of the total energy.

6 METHODS TO REDUCE CONFIGURATION ENERGY

The dynamic energy associated with programming the FPGA can be reduced by reducing the capacitive load and by reducing the number of writes required to program the FPGA. The total capacitive load is determined by the parasitic capacitance from the metal wires. This is determined by the physical size of the array and is fixed. It is possible to hierarchically isolate the tiles to reduce the switched capacitance. The number of writes is dependent on the number of configuration storage cells that have to be programmed, and can be reduced by reducing the number of storage cells. Since dynamic energy is a function of the number of times a capacitor is charged and discharged, controlling the transition activity can reduce the dynamic energy.

6.1 Selective Tile Activation

In the preliminary implementation shown in Fig. 3, the configuration data were distributed over the entire array. This can result in unnecessary switching of capacitance. Using methods similar to memory design, selective activation of blocks can be used to reduce energy. In the case of the FPGA, only the configuration distribution circuitry of the target tile needs to be activated as shown in Fig. 4. This has a dramatic impact on the local energy.

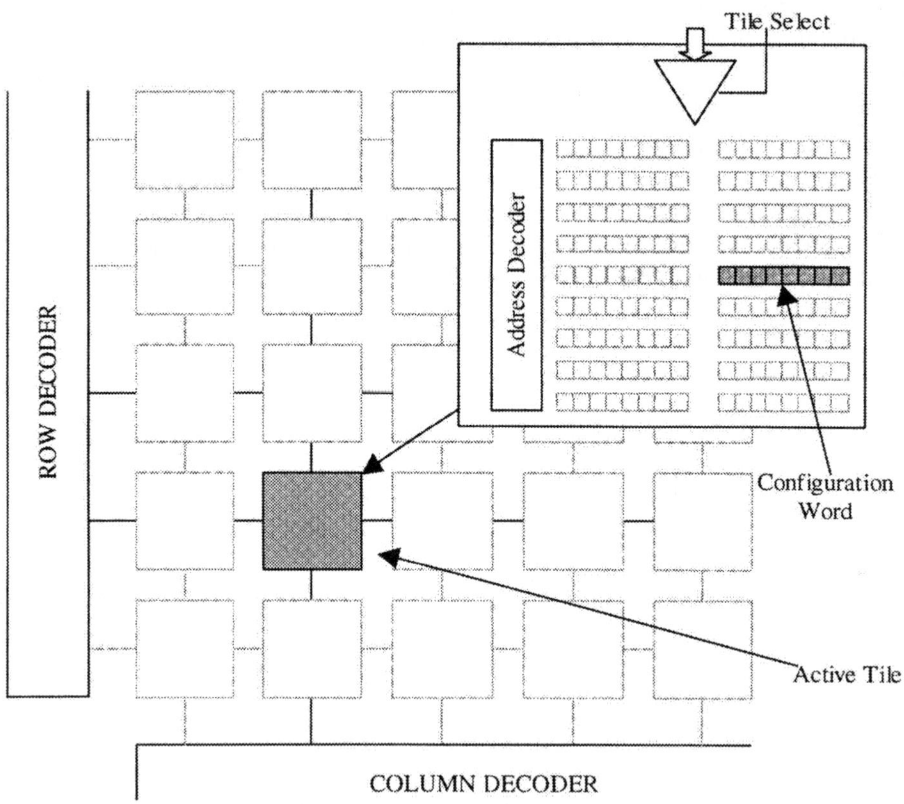

Figure 4. Selective Tile Activation

6.2 Configuration Compression

The number of cycles that are required to program an FPGA is dependent on the number of memory cells used to store the configuration. In traditional FPGAs, there is a one-to-one correspondence between the programming switches and the memory cells used to control them. In some cases, it is possible to reduce the number of memory cells required to store the configuration.

As an example, the connection boxes are implemented using pass transistors with one configuration bit per transistor, as shown in Fig. 5. This provides precise control over each switch, and is the way connection boxes are commonly implemented in FPGAs. The control of each switch independent of the other switches is required in the connection boxes of output pins to facilitate fan-out at these nodes. This is not necessary for the connection boxes of the input pins where only one source will be driving the input pin at any given time.

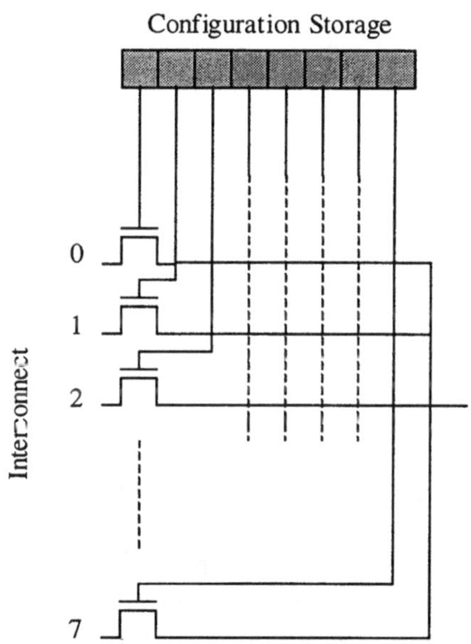

Figure 5. Conventional Implementation of Connection Box

This property of the input connection box can be taken advantage of to reduce the number of memory cells. One method would be to implement the connection boxes using multiplexers as shown in Fig. 6(a). Even though the number of memory cells is reduced, the input signal will see a chain of pass transistors. This will adversely affect the performance of the FPGA, and is undesirable.

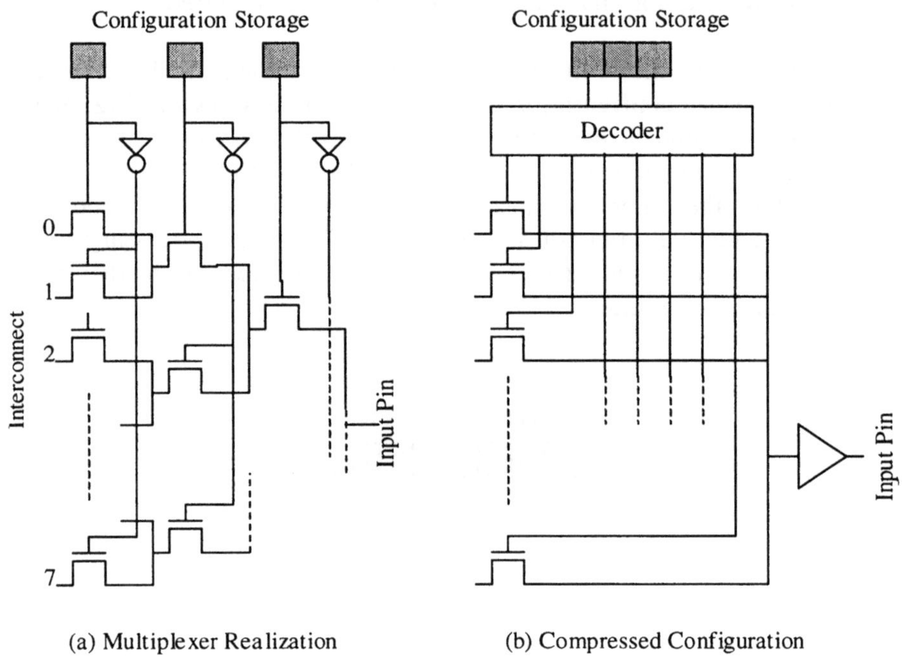

Figure 6. Configuration Compression

The second method is to binary-encode the configuration for storage as shown in Fig. 6(b). The reduction in the number of memory cells used to store the configuration bits necessitates the use of a decoder to expand the configuration. The area impact of using an encoder has to be evaluated, since configuration storage dominates the area of the array. In Section 5 it was pointed out that the energy contribution of the memory cell being programmed is less than 0.5% of the total energy. Hence, the energy consumption of the decoder can be ignored unless it increases the energy of storage by a few orders of magnitude. During the configuration step, the delay in the critical path is just the time required to latch the data in the storage cells. Hence, the delay overhead of the decoder can be ignored.

The overhead of using the encoding technique is shown in Fig. 7. The plots are of the ratio between the area (energy) of the encoded method and the area (energy) of the un-encoded configuration storage. The x-axis is the number of number of memory cells that are being grouped for binary encoding.

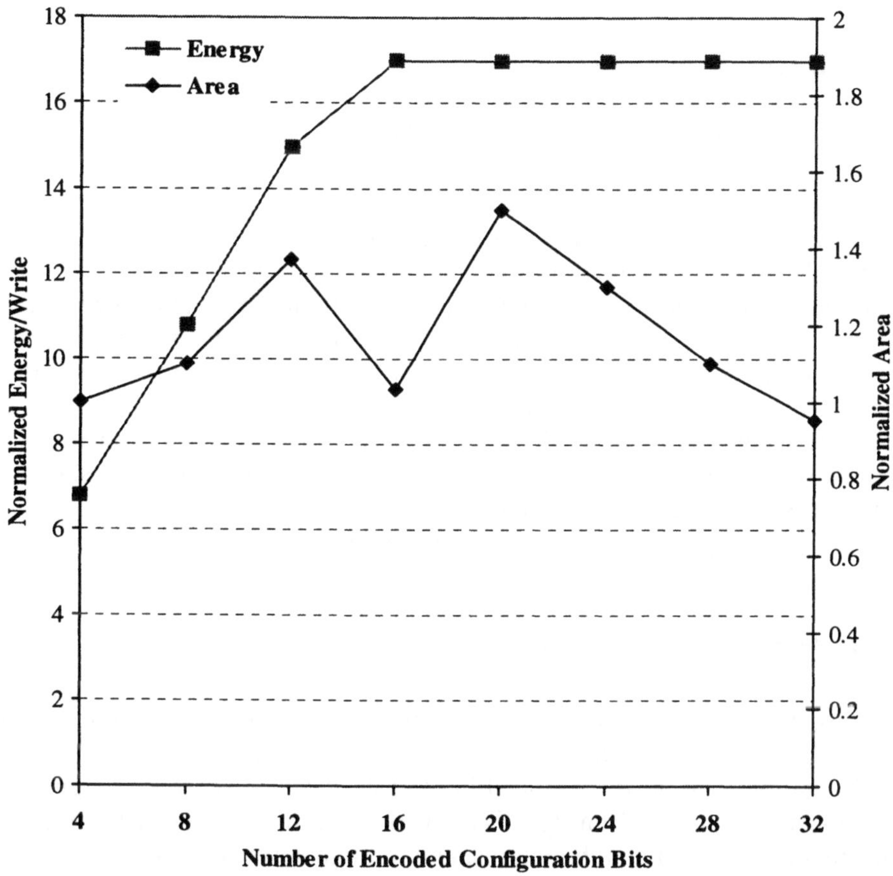

Figure 7. Effect of Encoding Configuration Bits

It is seen that the area overhead is minimal if the cells are combined properly. Although the energy overhead looks prohibitive, 7 to 17 times more expensive, this affects only the storage energy. This component was shown to be barely 0.5% of the total energy. The rest of the configuration energy (99.5%) reduces proportionally with the reduction in the configuration space.

By using this method of compression, the configuration space can be reduced by ~20% for the low-energy FPGA architecture. Judicious arrangement of the routing architecture can also make more bits available for this encoding technique.

6.3 Software Technique

Dynamic energy in a given static CMOS circuit is dependent on the transition activity. In most cases, this activity is dependent on the sequence of data being processed, and is beyond the control of the designer. In the case of the FPGA, the statistics of the bits to be loaded onto the configuration space of the FPGA are available prior to the configuration step.

In the random access technique, each configuration data word has an address associated with it, pointing to the resource being programmed. Therefore, the ordering of the words in the configuration step will not affect the implementation of the application. This characteristic is exploited by the software technique to order the write cycles to minimize the transitions on the configuration bus. The effect of this software method will be shown in Chapter 8. This technique cannot be employed in the shift register method, where the configuration bits cannot be rearranged.

7 CONCLUSION

Using the FPGA as a true embedded performance accelerator block demands frequent programming of the array. In this usage model, the configuration overhead of the FPGA can easily become a significant component of the total energy. The overhead can be reduced by adopting different programming architectures, and through software techniques.

The shift register method employed in commercial FPGA architectures is not suitable in the new role of the FPGA. The random access technique is advantageous from a delay and energy perspective if the FPGA is going to be programmed frequently.

Based on a preliminary implementation of the random access method, the different components of energy are isolated. Based on this, different modifications are advocated to further improve the basic implementation.

Hierarchical decoding of the address space is done to minimize the capacitance switched during each write cycle. The capacitive load and the configuration space are reduced by encoding the configuration. The configuration space is reduced by almost 20% using this technique.

Software methods can be used to minimize the transition on the address and configuration data busses. This is made possible by the fact that in the random access method, the ordering of the configuration word does not have any effect on the final implementation.

Chapter 7

HARDWARE IMPLEMENTATION

1 INTRODUCTION

A physical implementation is invaluable to completely verify the proposed architectural and circuit level techniques. A stand-alone FPGA of 256 logic blocks with an equivalent logic capacity of 512 4-input lookup table is implemented. This prototype, LP_PGAII, is useful for exhaustively testing the different architectural and circuit features. One of the principal functions envisioned for the low-energy FPGA is as a performance accelerator used in conjunction with dedicated functional units for low energy applications. A smaller version of the prototype array is embedded in a heterogeneous processor for base-band coding applications.

This chapter describes the hardware implementation details of the different components that make up the FPGA. The implementation of the logic block, connection boxes, interconnect levels, and the configuration architecture are discussed. The design takes into consideration the effect it has on the energy and the programming overhead.

The physical implementation is in a 0.25μm CMOS process, from STMicroelectronics. The process provides a poly and six metal layers for a standard digital process.

2 LOGIC BLOCK

A logic block structure capable of implementing a 5-input random logic function or a 2-bit arithmetic function was shown to be optimal for energy efficiency in Chapter 4. This functionality is made possible by implementing the logic block as a cluster of 3-input lookup tables. This clustering technique makes it possible to combine the results of the four 3-input LUTs in various ways to simultaneously realize up to three different functions in a logic block. The combination of the results of the 3-input LUTs is realized using multiplexers that can be programmed at time of configuration.

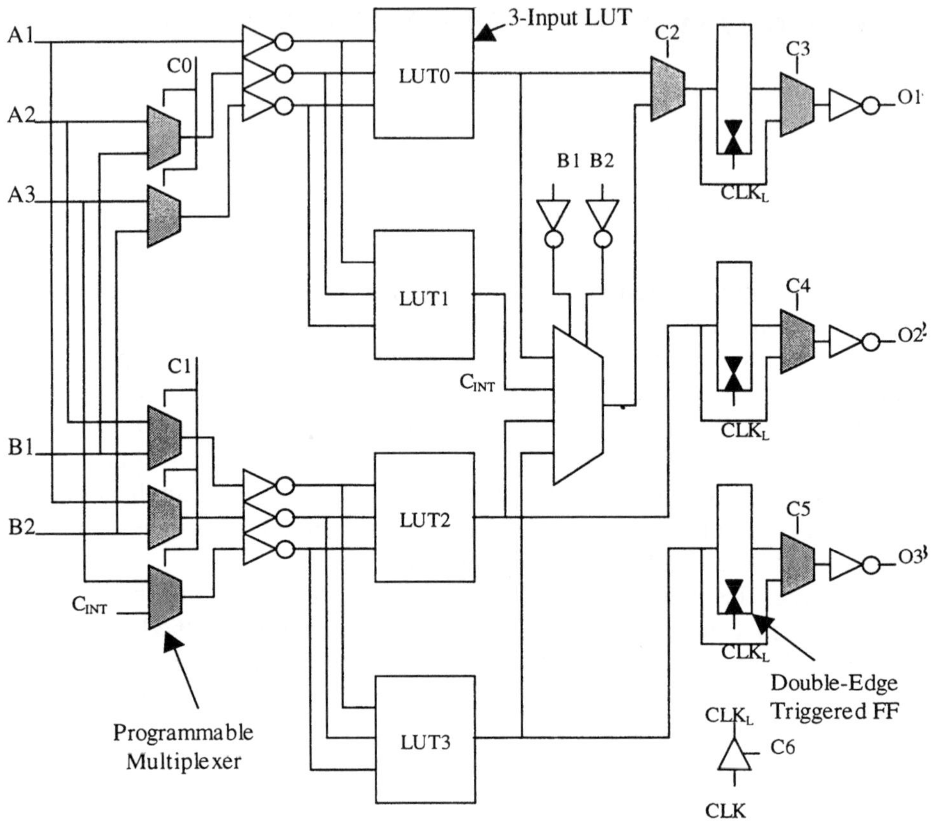

Figure 1. Logic Block Implementation

All three outputs of the logic block can be registered if required. The flip-flops use double-edge-triggered clocks to reduce the clock activity on the clock distribution network for a given data-throughput. The block diagram of the logic block is given in Fig. 1. The multiplexers that can be programmed at the time of configuration are shown as shaded.

2.1 Lookup table

The 3-input lookup table that is used in the logic block is implemented using a multiplexer as shown in Fig. 2.

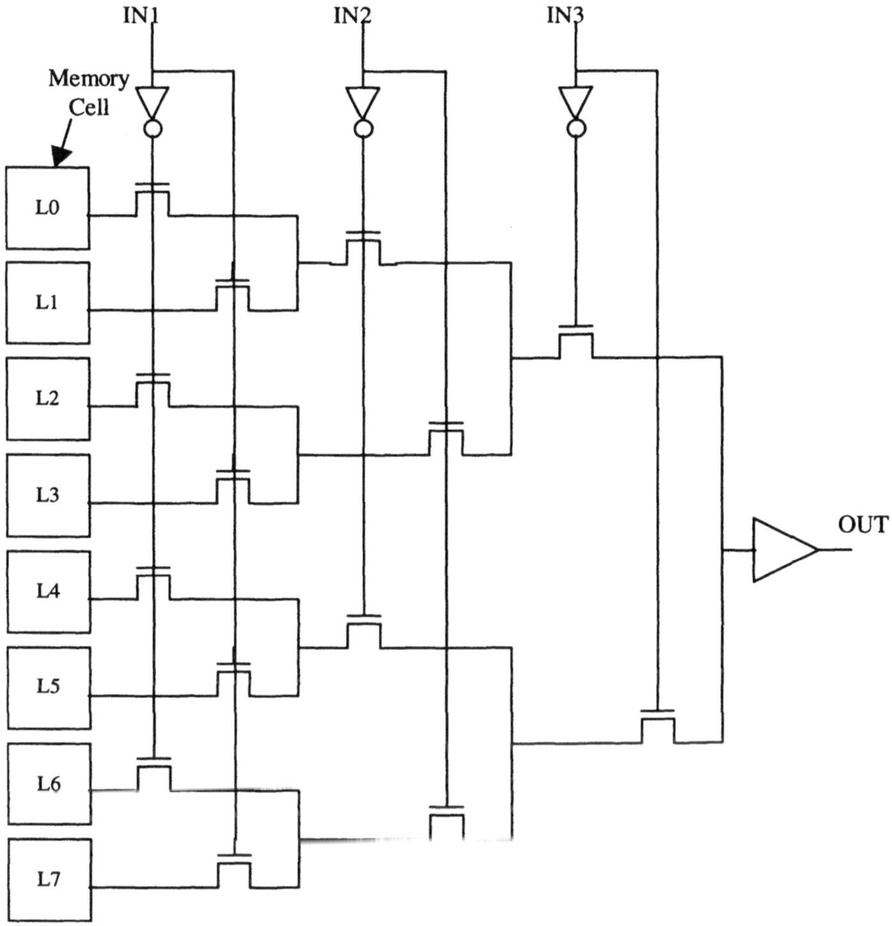

Figure 2. Lookup Table

The control signals of the multiplexer are the inputs to the LUT. The inputs to the multiplexer are stored in memory cells. The functionality of the LUT is controlled by programming the contents of the memory cells, L0-L7, based on the truth table of the required function.

2.2 Logic Block Programming

The functionality of the logic block can be controlled by programming the multiplexers, and the contents of the lookup table.

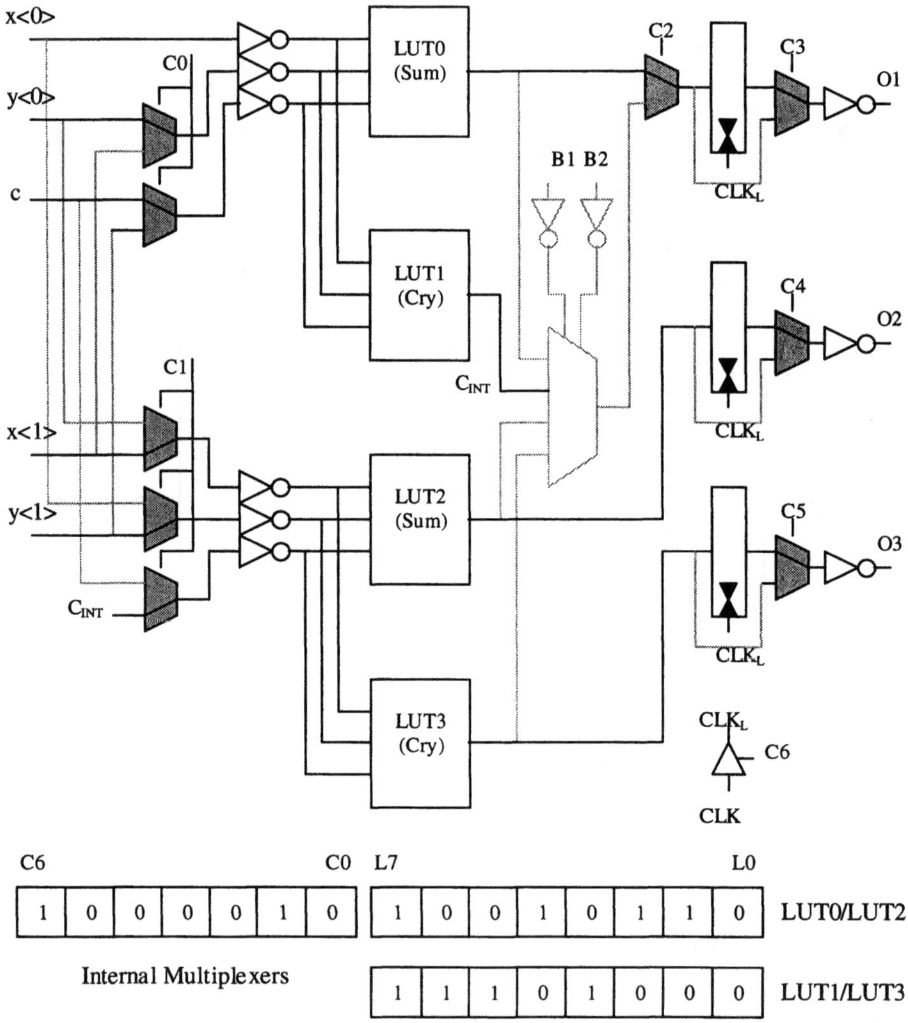

Figure 3. Logic Block Programmed for 2-Bit Addition

Configuration bits C0 and C1 control the data being fed into the LUTs. Configuration bits C3, C4, and C5 are used to control whether the outputs are registered. C6 is used to enable the clock distribution inside the logic block. If the flip-flops are not active, disabling the internal distribution of the clock can reduce the clock energy considerably.

Fig. 3 shows how the logic block can be programmed to implement an addition operation with operands X<1:0>, Y<1:0> and a carry in bit (C). The outputs are Z<1:0> and the carry-out bit (Cout).

2.3 Logic Block Delay

The critical path exercised in the logic block is dependent on the function being realized. The delay for implementing a clocked 5-input random function is 4.5nS. The delay for implementing a two bit, clocked arithmetic function is 5nS.

3 INTERCONNECT

All three levels of the interconnect hierarchy are implemented in the low-energy FPGA. The realization of the interconnect primitives is dependent on the exact implementation of the interconnect architecture.

3.1 Interconnect Levels

3.1.1 Nearest Neighbor Connection (Level-0)

The Level-0 connections provide connections between adjacent logic blocks. Each output pin connects to one input pin of the eight immediate neighbors. The output-input connectivity pairs are O1-A1, O2-B1 and O3-A3. This means that the output pins O1, O2, and O3 can connect to the input pins A1, B1, and A2, respectively, of each of the neighboring eight logic blocks.

Figure 4. Nearest Neighbor Connections

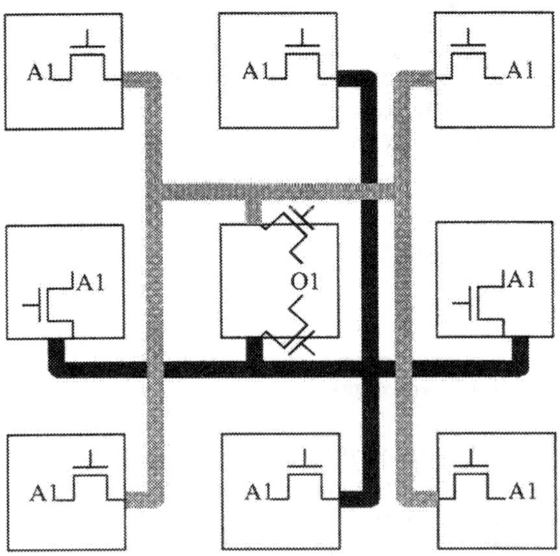

Figure 5. Level-0 Implementation

The routing overhead of having eight separate lines to each input pin from the output pins of the neighbors is quite high. The overhead can be reduced if multiple pins share the same interconnect line. Fig. 5 illustrates this concept for the O1-A1 connection. There are two global lines per input pin. Each of the output pins of the adjacent logic blocks is connected to these global lines through switches. To realize a connection between an output pin and an input pin, the two switches connecting the two pins to the global line must be enabled.

3.1.2 Mesh Architecture (Level-1)

The Mesh Architecture is realized with a channel width of five. The pins of the logic block are uniformly distributed on all sides of the logic block. The pins of the logic block can access all tracks in the corresponding routing channel. The switch box allows connections between each routing segment in a given channel and the corresponding segments in the other three routing channels. The block diagram is as shown in Fig. 6.

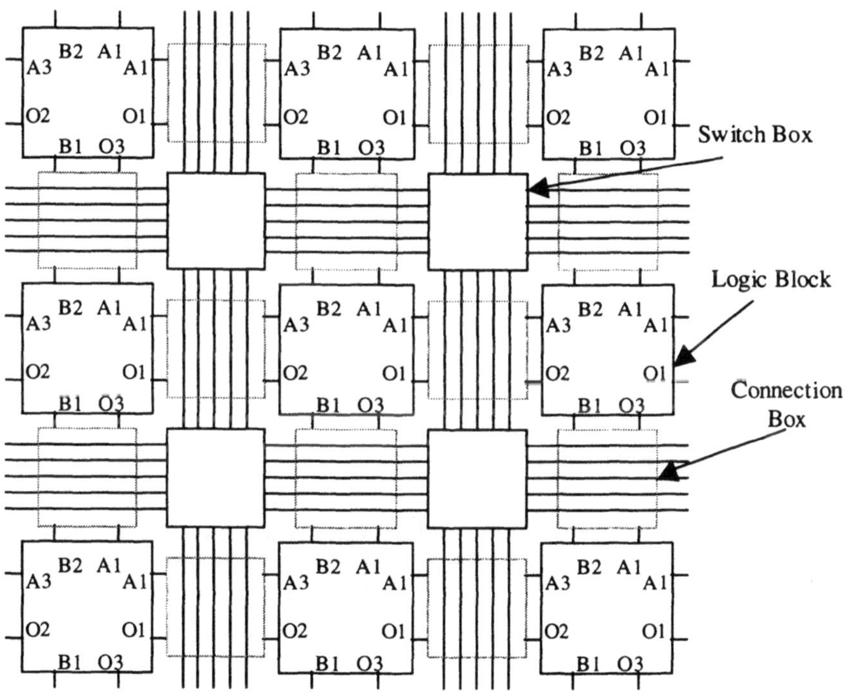

Figure 6. Level-1 Connections

3.1.3 Inverse Clustered Tree (Level-2)

The Level-2 network provides connection between logic blocks that are farther apart on the array. The long connection can be accessed through the Mesh structure. Two tracks in each routing channel are connected using the Level-2 network. This is illustrated in Fig. 7. The routing through the different levels of the Level-2 network is realized using the 3-transistor routing switch shown.

During the physical implementation, the Level-2 network contributes a significant amount to the area. Area minimization can be achieved by recognizing that the higher levels of the network can be discarded without any significant penalty to the routability.

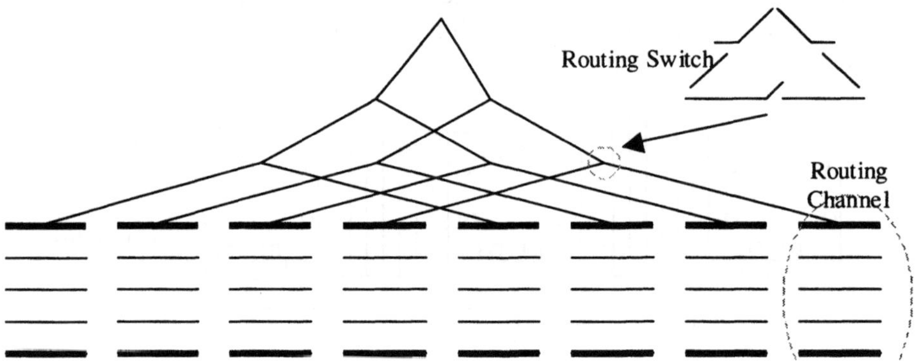

Figure 7. Level-2 Connections

3.2 Interconnect Primitives

The connectivity of the routing architecture is achieved using connection boxes and switch boxes. The connection box architecture determines the connectivity of the input and output pins of the logic block to the general purpose routing. The switch box provides the connections between the routing channels.

3.2.1 Connection Box

The connection box controls connectivity between the pins of the logic block, the tracks in the routing channel, and the Level-0 connections. The connectivity of the channel is one, while the size of the neighborhood for the Level-0 connections is eight. The connection box for the pins is realized using pass transistors. Fig. 8 shows the input connection box for the pins that connect only to the routing channel.

If an input pin is unused, the input to the receiver is floating. Energy dissipation in the form of short circuit current in the receiver can occur in such situations. This is prevented using the NMOS transistor connected to the ground. The transistor is activated if the programming of the connection box indicates that the input pin is unused. This does not require another configuration bit

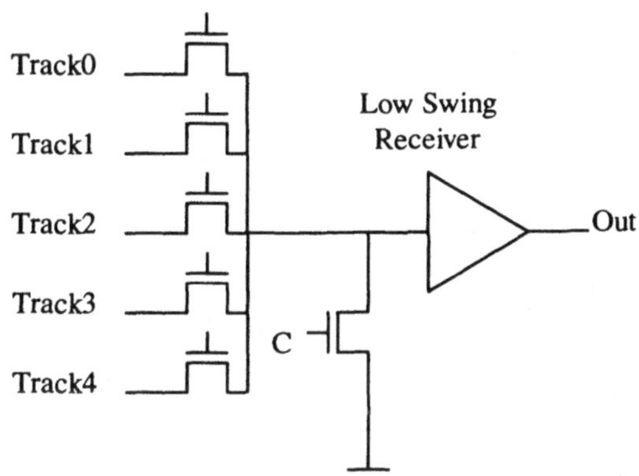

Figure 8. Type-A Connection Box

Different implementations are possible for the connection boxes for the pins that can access both the Level-1 routing channel and the Level-0 network. Fig. 9 illustrates two implementations of the input and output connection boxes for a channel width of five and a neighborhood size of eight. Both implementations are comparable from an energy and delay perspective.

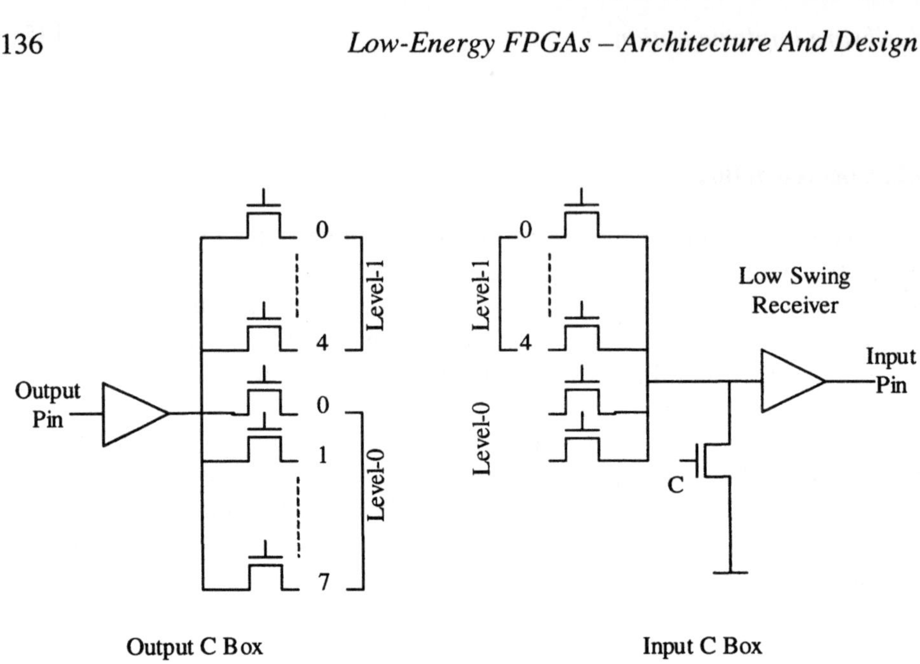

(a) Level-0 Control At The Output C Box

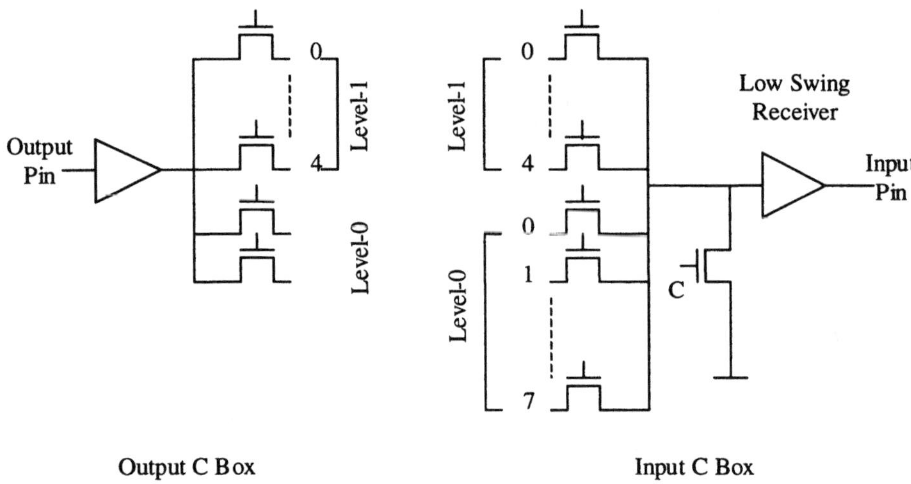

(a) Level-0 Control At The Input C Box

Figure 9. Connection Box for Input Pins

The implementations shown in Fig. 9 differ in the configuration overhead. If no encoding of the configuration is done, both implementations require twenty bits to program them completely. It was discussed in Chapter 7 that the configuration bits for the input connection box can be encoded since at any given time only one connection is possible. If the encoding technique is used, the first implementation requires sixteen bits for programming. By moving the control of the Level-0 connections to the input connections box, more configuration bits are available for encoding. For the second implementation, only eleven bits are required.

This represents a 45% reduction of configuration size over the traditional programming. This illustrates how seemingly equivalent implementations can be exploited to improve the configuration size.

3.2.2 Switch Box

The switch box used in the low energy FPGA has a flexibility of three. Each track can connect to the corresponding tracks on the other three sides. The switch box connectivity is shown in Fig. 10. The connectivity in the switch box is realized using 6-transistor structures for each track.

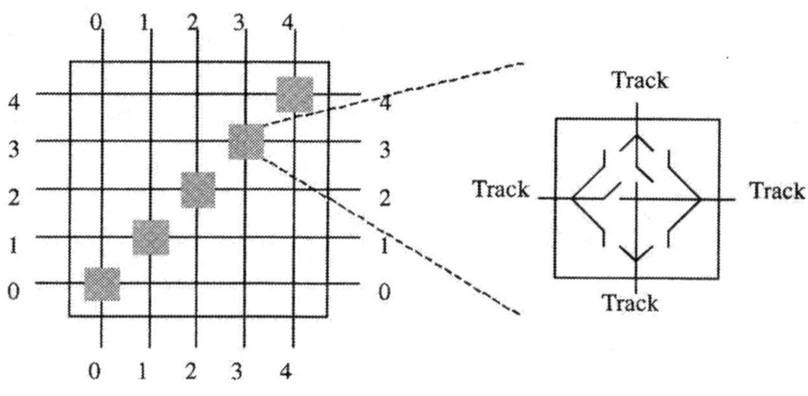

Figure 10. Switch Box

4 TILE LAYOUT

The logic block, connection boxes, and the switch box have been combined to form a single tile. The tiles can be used to create arrays of different sizes by abutting the tiles. The layout of a single tile is shown in Fig. 11. The dimensions of the tile are 241μ x 219 μ in a 0.25 μ process.

The contribution of the different components to the total area is given in Table 1. The routing resources account for approximately 49% of the total area. As the size of the array increases, the fraction of the total area used by the routing will also increase. This is because the increase in the array size necessitates an increase in the routing resources required for each tile to ensure successful routing. The logic block contributes only 9% to the total tile area even for such a small array.

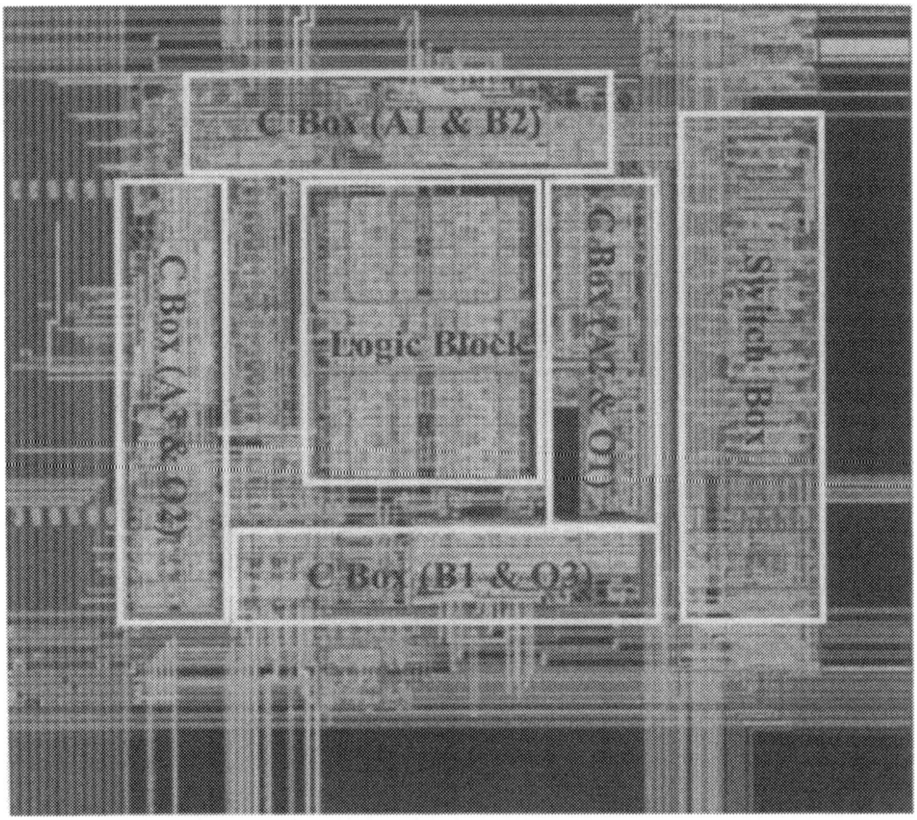

Figure 11. Layout of a Tile

Table 1. Contribution of Different Components to the Total Area

Component	Percentage of Total Area
Logic Block	9
Connection Box	18
Switch Box	10
Hierarchical Routing	21
Local Configuration Distribution and Address Decode	5
Global Configuration Distribution	13
Miscellaneous Routing	24

5 CONFIGURATION ARCHITECTURE

The configuration method used in the low-energy FPGA is that of a random access technique. This makes it possible to selectively program the resources in the FPGA, without having to program the entire array each time. The configuration architecture discussed in Chapter 7 is repeated in Fig. 12.

The configuration bits in each tile are grouped as 8-bit words. Related bits are grouped to form a word. For example, all the bits required to program the connection box of a specific logic pin are grouped together. This helps to minimize the number of writes required to program each specific resource. Internally to each tile, a 5-bit address is used to select the configuration words. Global row and column decoders are used to select the tile that is being programmed.

The physical layout of the tiles is such that each column presents an 8-bit configuration data bus. At the global level, these busses can be combined to form an n-Byte configuration data bus. This can reduce the number of writes required to program the entire array. For example, if the busses are combined to form a 32-bit bus, then during each write, the same resource in each of four tiles can be programmed. This can potentially reduce the number of writes by a factor of four. Often the reduction in the number of writes is less than this best-case scenario.

The configuration cells can only be written into, the contents cannot be read back. In this scheme, it is sufficient to have a simple latch to store the configuration. If traditional serial programming is used, a master-slave flip-flop has to be used. In terms of area, a master-slave flip-flop is more expensive than a simple latch. The circuit of the latch is as shown in Fig. 13.

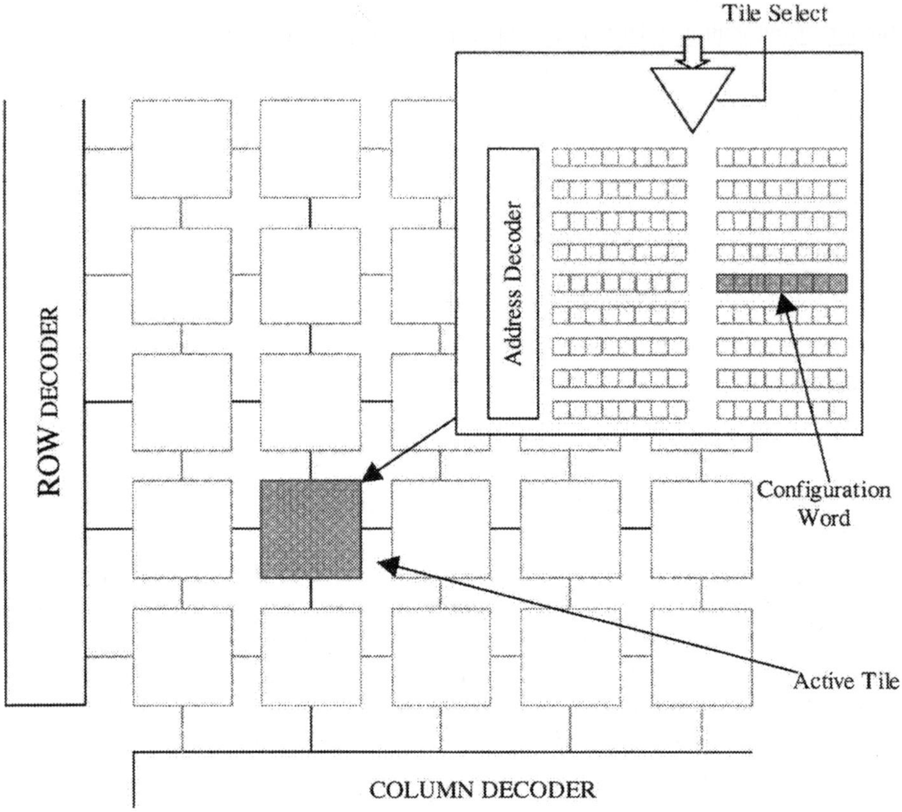

Figure 12. Random Access Programming

The storage cells are provided with a reset mechanism to disable the routing switch to which it is connected. During start-up, the reset is enabled to bring all the configuration bits to a known state. This prevents the short-circuit currents that can occur when multiple output pins are connected together.

Another advantage of initializing the configuration bits is the reduction of programming required. Consider the switch box; the connections in the switch box will not affect the short-circuit current. However, when an application is mapped into the array, it has to be ensured that there are no unwanted connections in the switch-box. This requires additional writes. If the bits in the switch box are initialized to a known state using the global reset mechanism, these extra writes can be avoided.

The programming speed is determined by the delay of the address decoder. For the prototype chip of size 16 x 16, Fig. 14 illustrates the delay components. The worst-case delay of the global decoder is obtained by loading the global decoder with a metal trace running the entire length of the array. For the local decoder, the load used is that of a line running the length of the logic block. If each resource in the array needs to be programmed, ~5700 configuration write cycles are required. For most applications, the number of cycles is significantly lower.

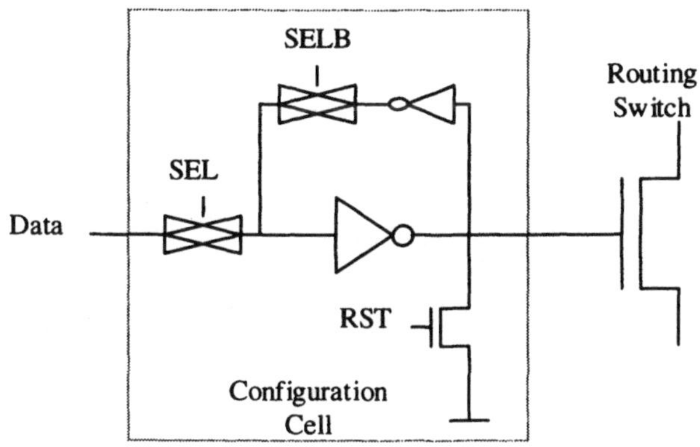

Figure 13. Configuration Storage

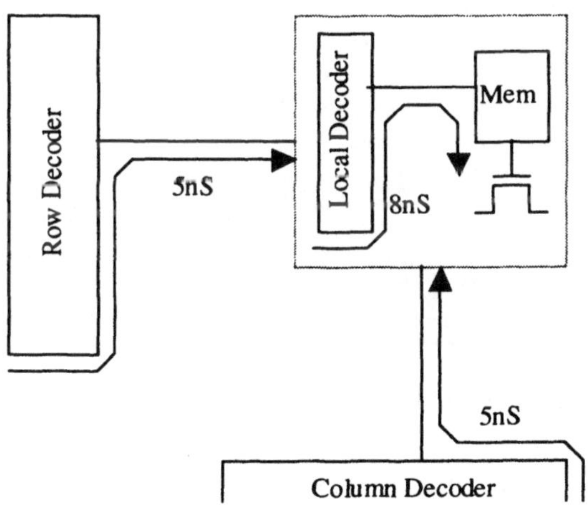

Figure 14. Delay Component of the Configuration Step

6 FINAL LAYOUT

The final layout of the low-energy FPGA, LP_PGAII, is given in Fig. 15.

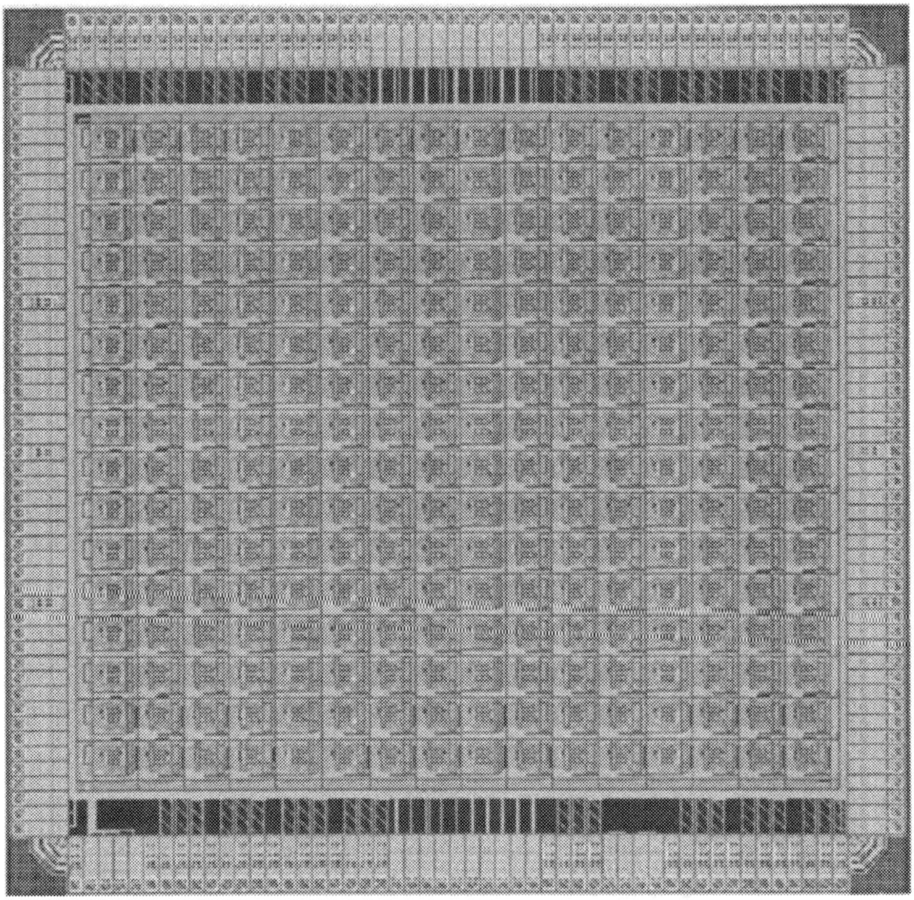

Figure 15. Chip Micrograph of LP_PGAII

The chip is an array of size 16 x 16, with an equivalent logic capacity of 512 4-input lookup tables. Three separate power supplies are routed to the chip. A 1.5V supply is used for the input/output pads, and 1.5V/0.8V for the FPGA core. The lower voltage is for the low-swing circuit. The chip measures 4.7mm x 4.7mm in a 0.25μm process. The contribution of the array is 4mm x 4mm. The rest of the area is from the pad ring. Table 2 summarizes the details of the chip.

Table 2. Chip Specifications

Array Size	16 x 16
Power Supply	1.5V and 0.8V
Chip Size	4.7mm x 4.7mm
Core Array Size	4mm x 4mm
Process	0.25μm CMOS

7 FPGA AS AN EMBEDDED UNIT

One of the main applications envisioned for the FPGA is that of an embedded performance accelerator block. The Pleiades heterogeneous architecture [Abnous98] is an ideal environment to use an embedded FPGA.

7.1 Pleiades: An Introduction

The key to achieving power and performance efficiency is to match algorithmic constructs with the appropriate architectures. In wireless embedded systems, the algorithms of interest include high level control-flow and dataflow-intensive operations. While control-flow constructs are best implemented in programmable proccssors, dataflow computation benefits greatly from application-specific datapaths. The Pleiades reconfigurable architecture achieves low energy consumption by providing a computational platform comprising of heterogeneous functional blocks (i.e. microprocessor, ASIC modules, FPGA). In addition, since many of the dataflows perform computations using basic computational modules (MAC, MEM access), providing custom ASIC for each dataflow computation is not necessary. Providing reconfigurability based on the basic computational modules gives flexibility and energy efficiency.

The Pleiades architecture is composed of a programmable microprocessor and heterogeneous computing elements (referred to as satellites) connected via a reconfigurable interconnect network. The architecture template is shown in Fig. 16. In addition, the architecture template fixes the communication primitives between the microprocessor and satellites and between each satellite. The template allows domain specific architecture instantiation. For each algorithm domain (e.g. communication, speech coding, image and video coding), an architecture instance can be derived with the required type and number of satellites.

Within the Pleiades architecture, the embedded microprocessor does the high–level control and spawns off intensive computations to the satellite processors.

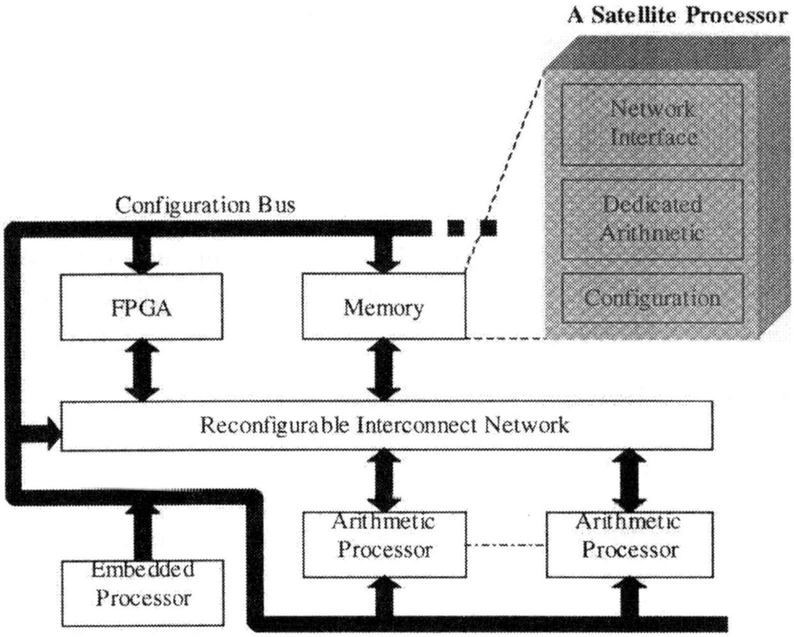

Figure 16. Heterogeneous Architecture Template [Abnous98]

Architectural optimization for satellites is important since a high percentage (70-90%) of the entire application computation power needs to be accelerated by satellites. To reduce overhead in terms of instruction fetch and global control, the satellite architecture utilizes distributed control and configuration. The programmer can specify basic satellite configurations and configurations for the reconfigurable interconnect to build a cluster of satellites. To achieve distributed control, each satellite is equipped with an interface that enables it to exchange data streams with other satellites efficiently, without the need for a global controller. The communication mechanism between each satellite is data driven.

7.2 FPGA Satellite

One of the satellites used in the Pleiades architecture is an FPGA. The FPGA design described in this book fits the Pleiades target of low energy consumption.

The FPGA satellite embedded in Pleiades is a 4 x 9 array of logic blocks. This fits the requirements in terms of the logic capacity. The logic block array used employs the architectural and circuit level features discussed in the previous chapters. The inputs and outputs of the array are arranged as two 16-bit input ports, and one 16-bit output port. The configuration loading is via a 32-bit configuration bus. Data transfer between ports of different satellites is achieved using a two-phase handshake protocol.

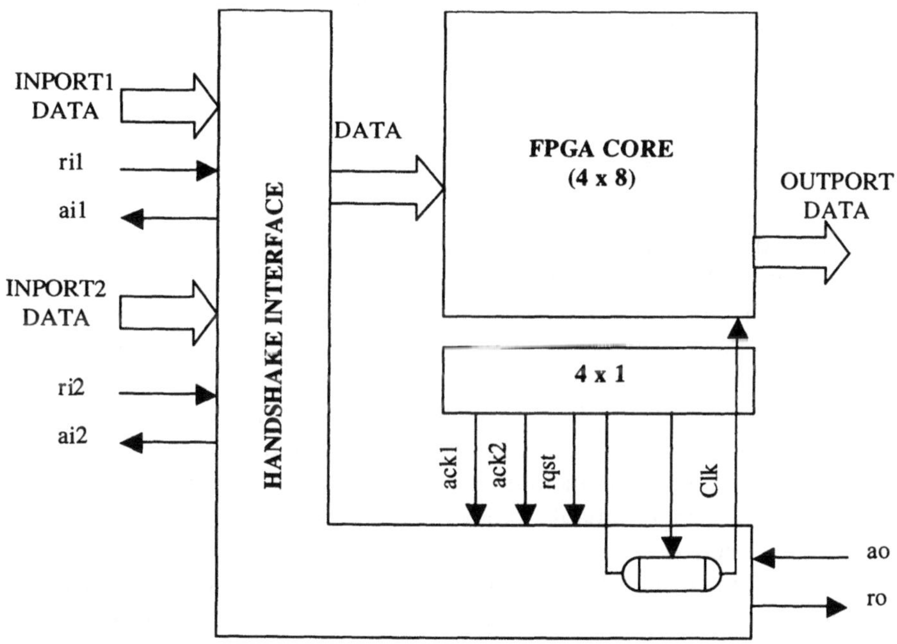

Figure 17. Satellite Block Diagram

7.3 Handshake Control

The FPGA satellite communicates with the other satellites using a two-phase asynchronous protocol. To enable this, each data bus is bundled with two handshake signals. An output port signals the availability of data by activating the request token, *ro*. When an input port detects the availability of new data, it latches the data, and acknowledges receipt using the signals *ai*. This protocol at the periphery of the FPGA core is realized using the handshake interface. The functionality and implementation of the handshake interface is described in [Benes99].

For a satellite unit with a specific functionality, the handshake signals can be generated using fixed logic. This is because the type of data (vector/scalar) and the pipeline depth of the functional block are fixed. In an FPGA satellite, the data type and the pipeline depth are dependent on the function being implemented. Hence, the generation of the handshake signals has to be programmable. One of the options is to design an interface block so that it can be programmed to generate the handshake signals. In this work, it was decided to utilize the programmability of the FPGA to implement the handshake signals. In the FPGA satellite, a row of logic blocks is made available for generating the handshake control.

7.4 Programmable Clock

The Pleiades environment uses a Globally Asynchronous Locally Synchronous (GALS) methodology. Each satellite unit has its own internal clock, and is a synchronous island. The local clock is activated only when a new data-token arrives, and the period is dependent on the logic depth.

For a satellite with fixed functionality, the clock frequency is fixed. If a fixed clock is used in the FPGA satellite, the logic implementation in the FPGA has to conform to the fixed clock. This can be quite inefficient. Rather than constraining the implementation of the functions to conform to a fixed clock period, a programmable delay element is incorporated to generate different clock periods.

The delay element can be programmed to obtain a delay of one to eight times the unit delay. At the time of implementation, the delay element can be programmed to achieve the clock frequency dictated by the logic depth and the critical path. This allows the fine-tuning of the local clock to extract the maximum speed out of the FPGA.

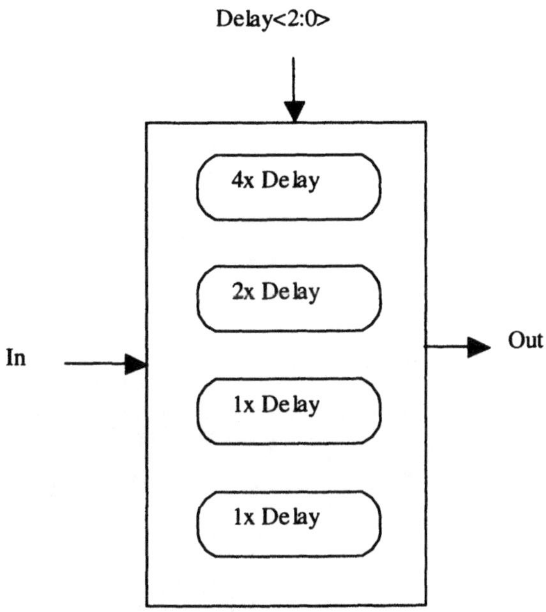

Figure 18. Programmable Delay Element

7.5 Layout

The final layout of the heterogeneous architecture, Maia, for voice band processing is shown in Fig 19. This processor combines an ARM8 processor [Burd00], dedicated functional units, and an embedded FPGA. The ARM processor spawns off compute-intensive tasks onto the dedicated functional units and the FPGA to realize speed and energy improvements. The MAIA architecture demonstrated more than an order of magnitude improvement in energy savings compared to a low-power DSP for voice coding applications [Zhang00a].

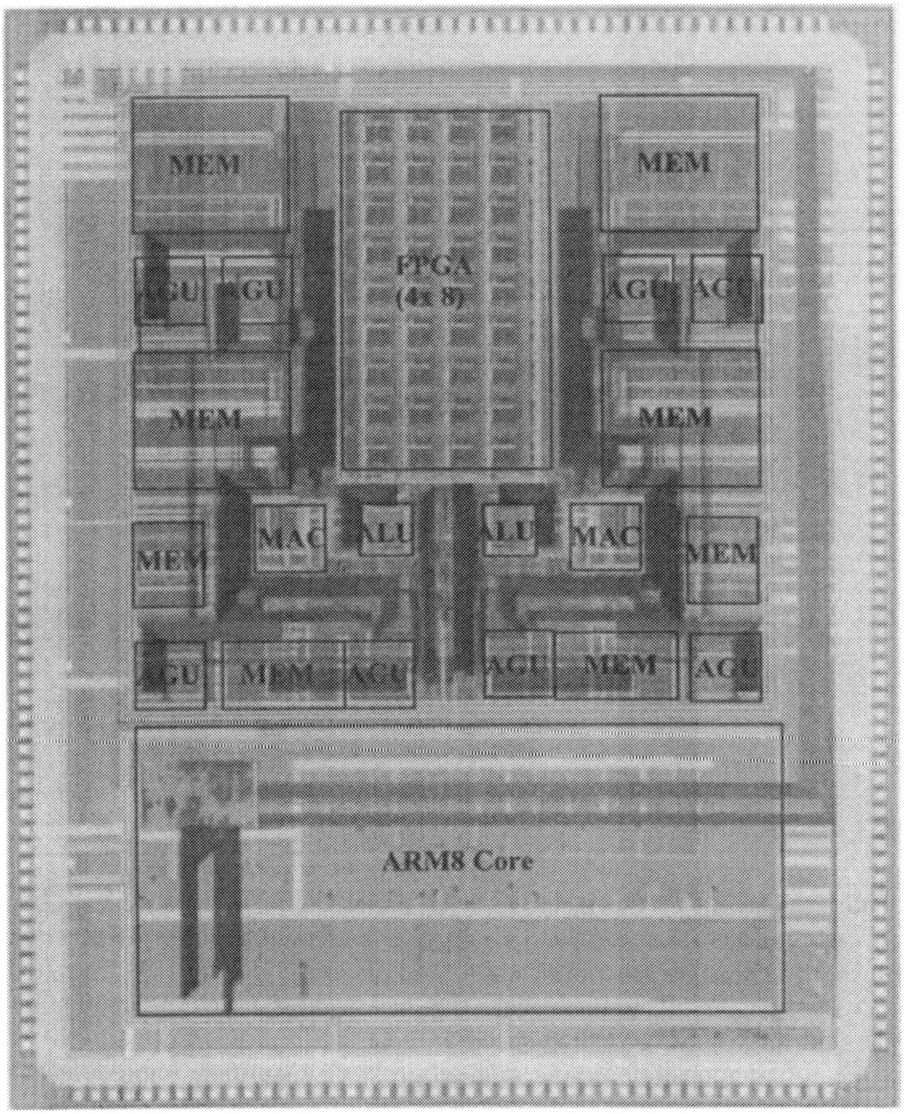

Figure 19. MAIA - Chip Micrograph

8 CONCLUSION

Physical implementation is one of the final steps in the validation of the design techniques and innovative architectures. This chapter describes the implementation details of two prototypes.

The first prototype, LP_PGAII, is a stand-alone array of 256 logic blocks with an equivalent logic capacity of 512 4-input LUTs. The specific implementation details of the critical components are described. The energy and performance impact of different implementations are evaluated.

The second prototype is a smaller array embedded in a heterogeneous reconfigurable architecture. In this version, the FPGA is used as an accelerator for implementing communication algorithms.

Chapter 8

RESULTS

1 INTRODUCTION

The verification of the architectural and circuit modifications is only complete after data are measured from a physical implementation. This chapter describes the measured data obtained from LP_PGAII. For comparison purposes, data from a commercial architecture are also reported.

The energy and speed performance of the FPGA is reported. This includes the configuration and execution steps of the implementation.

Measured data show a significant improvement in the energy efficiency of LP_PGAII when compared to commercial architectures.

2 MEASUREMENT SETUP

The basic measurement setup is shown in Fig. 1. The configuration and test data are loaded into the FPGA using the logic analyzer. The prototype board allows complete access to the different energy components.

2.1 Logic Analysis System

The HP 16702A logic analysis system is used for testing the prototype FPGA. The system is capable of generating test patterns at 100MHz and acquiring data at 200MHz. This is suitable for testing the prototype.

2.2 Prototype Board

The FPGA under test and the associated components are mounted on a prototype board. The output signal of the logic analysis system is TTL-compatible, and provides a 0-5V signal. The prototype has been designed to operate at 1.5V. Hence, a voltage level shifter is required to interface the logic analysis system to the FPGA. The chip used is a logic level down converter [74ALVC164245]. The down converter provides buffers that operate at the lower supply while being able to support input voltages that exceed the power supply.

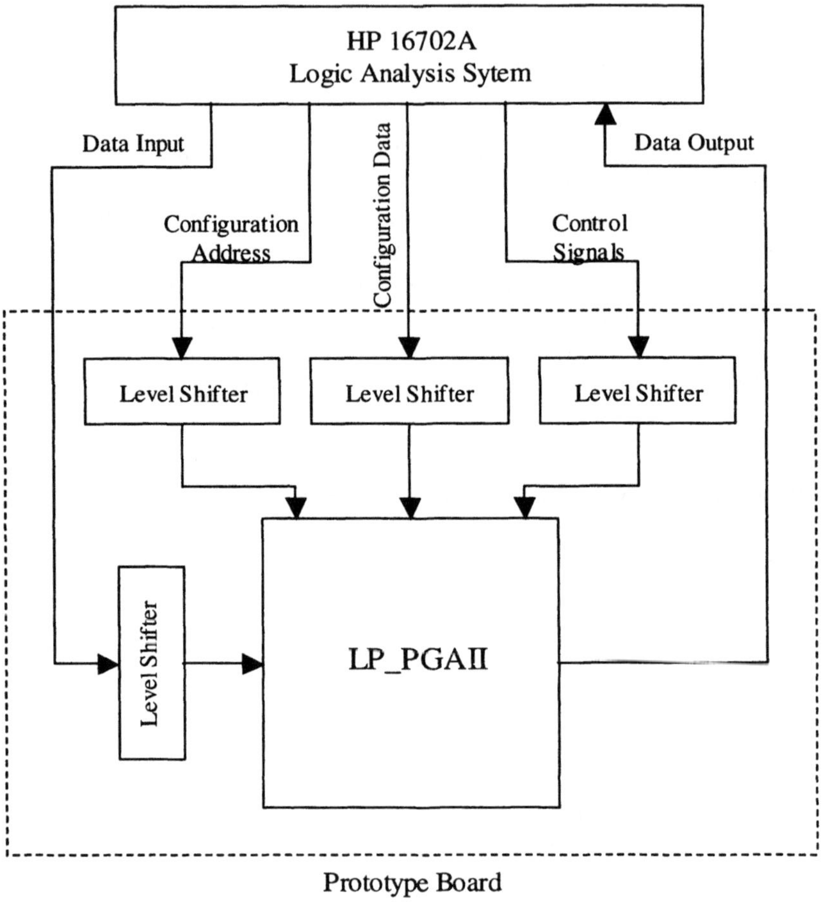

Figure 1. Block Diagram of the Test Setup

Separate power supplies are routed so that the power of the level converters, the IOs, and the core can be measured separately. This is required for a fair comparison. The prototype board is shown in Fig. 2.

Figure 2. Prototype Board

3 MEASUREMENT STRATEGY

A final comparison between FPGAs can be done only with measured performance data. The XC4000XL series from Xilinx is used as the representative commercial architecture. The XC4000XL is the low power series of the popular XC4000 architecture. To do a fair comparison of the architectures, the arrays have to be of similar logic capacity. This is to ensure that the larger arrays are not penalized due to the higher cost of the global signals. The XC4005XL is an array of 196 logic blocks with an equivalent logic capacity of 466 4-input LUTs. This is slightly smaller than LP_PGAII with an equivalent capacity of 512 4-input LUTs. The measurements are done using the XS40 prototyping board [XESS].

The XC4005XL is implemented in a 0.35μm process. The prototype chip is in a 0.25μm process. To do a fair comparison, the performance data measured from the XC4000XL FPGA have to be scaled for the 0.25μm process. Xilinx offers a new series of low-power FPGAs, XC4000XV, which has the same architecture as the XC4000XL, but is implemented in a 0.25μm process. This FPGA has a claimed power consumption that is lower than the XC4000XL by a factor of three, with a 30% improvement in speed performance. This factor for the energy and speed is used to scale the data obtained from the XC4005XL.

To compare the performance, benchmark circuits are implemented on LP_PGAII and XC4000XL. Identical data streams are input to the FPGAs to remove any data dependencies in the measurements. The energy of the architectures is compared for the same data throughput, rather than the same clock frequency. This is because LP_PGAII uses double-edge-triggering, and can support the same data throughput at half the clock frequency of the Xilinx architecture.

In a stand-alone FPGA, as is the case here, the energy dissipated by the IO pads cannot be ignored. The input pads are used to drive signals internal to the chip and their energy is relevant for comparison purposes. The output pads drive the metal traces on the printed circuit board and their energy consumption need to be considered to compare the architectures. Hence, the energy consumed in the output pads is subtracted out in all of the reported data.

4 MEASURED DATA

Measured data are reported in this section for LP_PGAII and XC4000XV. The speed performance of the FPGA is reported in terms of the toggle frequency. The energy of the FPGA is reported in two ways: the energy consumed in the interconnect as a function of length and the energy for implementing different applications. To get an idea of the cost of using these FPGAs in an environment where reconfiguration is important, both the energy and time for loading the configuration onto the FPGA are measured.

4.1 Toggle Frequency

One of the methods of specifying the speed performance of the FPGA is by the toggle frequency. The method is as shown in Fig. 3. A single output of the logic block is connected back to the input, and the delay of this loop is used as a measure of the FPGA performance. This delay includes the delays of the combinational logic, the flip-flop inside the programmable logic block, and that of the general-purpose routing.

The XC4000XV architecture is reported at a toggle frequency of 200MHz, giving a delay of 5nS. The LP_PGAII has a delay of 8nS, giving an equivalent frequency of 125MHz.

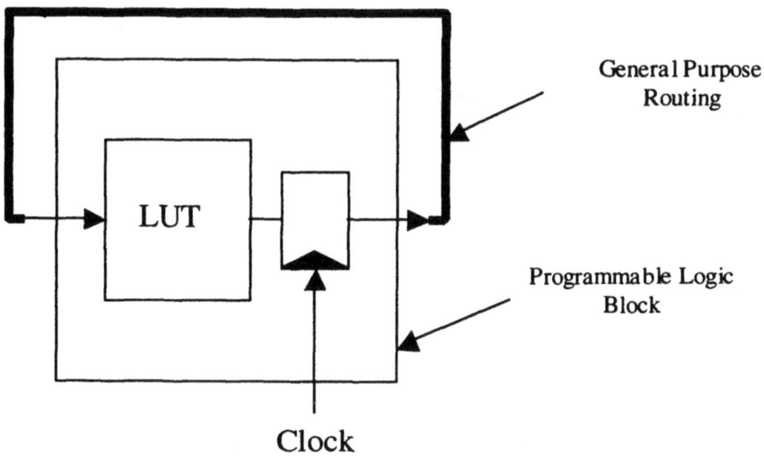

Figure 3. Set Up For Measuring Toggle Frequency

The toggle frequency is the maximum frequency that can be achieved. For most practical purposes, the frequency is much lower.

4.2 Path Length vs. Energy

Fig. 4 compares the energy dissipated in the interconnect for different path lengths. Since the logic blocks of XC4000XV and LP_PGAII are of similar logic capacity, the path lengths can be measured in terms of the Manhattan distance between the logic blocks. For comparison purposes, the best interconnect resource in each of the architectures is used to realize the path. For example, for the long paths, the Level-2 network in LP_PGAII and the long lines in XC4000XV are used. The energy contribution of the logic blocks is subtracted in the reported data.

The complete redesign of the interconnect architecture in LP_PGAII is reflected in the energy savings. LP_PGAII demonstrates energy reductions between ten to fifty times as compared to the Xilinx architecture.

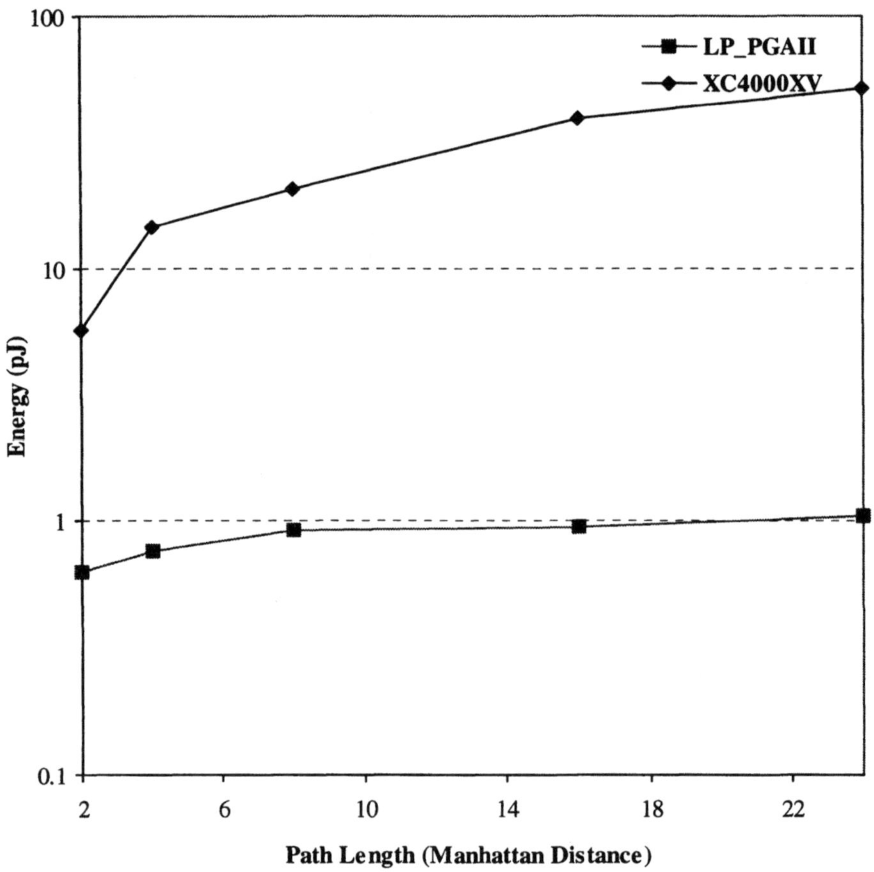

Figure 4. Energy as a Function of Path Length

In applications where the interconnect usage is substantial, the total energy will benefit considerably from the low-energy interconnect.

4.3 Execution Energy

To get a better idea of the overall energy consumption in the FPGA, applications are mapped onto the array and the energy is measured. The applications are described at the Register Transfer Level (RTL). The mapping to each FPGA is done based on the component library for the architecture. This helps to eliminate the energy variations that can occur if the two FPGAs use different RTL descriptions of the application.

The applications are executed with the same data throughput and input data streams. Since identical data streams are used, the dependency on the transition activity is taken into consideration. The energy is reported for processing one data token, and is given in Table 1.

Table 1. Execution Energy Per Data Token in pJ.

Application	XC4000XV	LP_PGAII
Single FF driving 9 segments	107	3.8
1 K Array of 16-bit counters	16667	750
Theta Function	183	20
Barrel Shifter	992	199
Accumulator	156	10
Viterbi Accelerator	1380	131

One of the common benchmarks used for comparing the power dissipation of FPGAs is the array of counters [Xilinx98][Altera97]. For this comparison, the energy dissipation of one flip-flop driving a 9 segment long interconnect is measured. A 1024 logic block array is filled with these elements configured as 16-bit counters. This gives a 12.5% activity factor on the interconnect.

This benchmark is interconnect-intensive due to the 9-segment load on each output pin. The low-energy interconnect in LP_PGAII exploits this advantage, and the energy is twenty-two times lower than the XC4000XV.

The other applications are typical functions that would be implemented in the FPGA to obtain performance acceleration. Reported energy is five to sixteen times lower than that of the commercial architecture. Larger improvements are obtained for functions that uses more interconnect resources.

4.4 Configuration Energy

The importance of the configuration energywas described in Chapter 6. When the FPGA is used as a performance accelerator, the cost of configuring the FPGA will have a strong influence on the functions that can be efficiently mapped onto the array. This section reports the energy for different applications with varying utilization of the array.

Table 2 compares the configuration overhead for programming the Xilinx FPGA and LP_PGAII. The configurations used cover a wide range of array utilizations. The "Route-2" function needs only two logic blocks to be programmed, while the "Viterbi Accelerator" uses ~70% of the array.

The speed overhead is reported in terms of cycles. For the Xilinx chip, the number of cycles is constant and independent of the application. This overhead does not consider the fact that while the Xilinx chip is programmed at 10MHz, LP_PGAII can be programmed at ~30MHz.

The difference in the configuration energy between the two FPGAs is dramatic, a reduction by three to six orders of magnitude. It can be seen that in LP_PGAII, the energy is a function of the utilization of the array, while the energy is constant for the Xilinx FPGA. The low configuration cost of LP_PGAII makes it a more attractive choice as a performance accelerator.

Table 2. Configuration Overhead

Application	Cycles		Energy (nJ)	
	XC4000XV	LP_PGAII	XC4000XV	LP_PGAII
Route-2	151910	14	8.6E6	3.2
Theta		191		46.1
Barrel Shifter		726		166
Accumulator		99		24.6
Viterbi Accelerator		1450		330

4.5 Configuration Reordering

The programming control provided by the random access technique can be used to control the transition activity during configuration. Before the configuration is written out, the order of the configuration address and data pair is arranged using a simple two-pass heuristic.

On the first pass, the configuration address-data pairs are ordered to exploit the locality of the configuration words. This means that the addresses to the same tile are grouped together. This ensures that the transition activity on the global row/column select lines is minimized. During the second pass, the writes within each address zone are ordered to minimize the transitions on the configuration data bus.

The improvement in configuration energy obtained by reordering the configuration writes is given in Table 3. The reduction in configuration energy is between 32% and 63%.

Table 3. Dependency of Configuration Energy on Ordering

Application	Random	Ordered
Theta	46.1	31.3
Barrel Shifter	166	61
Accumulator	24.6	9.9
Viterbi Accelerator	330	143

This heuristic is dependent on the configuration architecture and the process technology. This measurement is intended to illustrate the dependence of the configuration energy on the transition activity, and how the transition activity can be controlled in software to improve the energy performance.

4.6 Energy – Delay Tradeoff

One of the main goals of this low-energy FPGA project was to minimize the energy while maintaining acceptable speed performance. This resulted in using a 1.5V/0.8V power supply to achieve a maximum toggle frequency of 125MHz. It is possible to run this design at a higher voltage to improve the speed performance. Obviously, the improved speed can only be obtained at the cost of higher energy consumption.

Fig. 5 gives the relationship between the energy and speed for the accumulator application. The higher voltage supply was varied between 1.4V and 2.0V while the low voltage, which controls the swing on the interconnect, was varied between 0.8V and 1.1V. For each delay point, the high and low voltages were adjusted to obtain the lowest energy consumption. The speed performance of LP_PGAII can be improved by ~4x for an energy increase of ~80%.

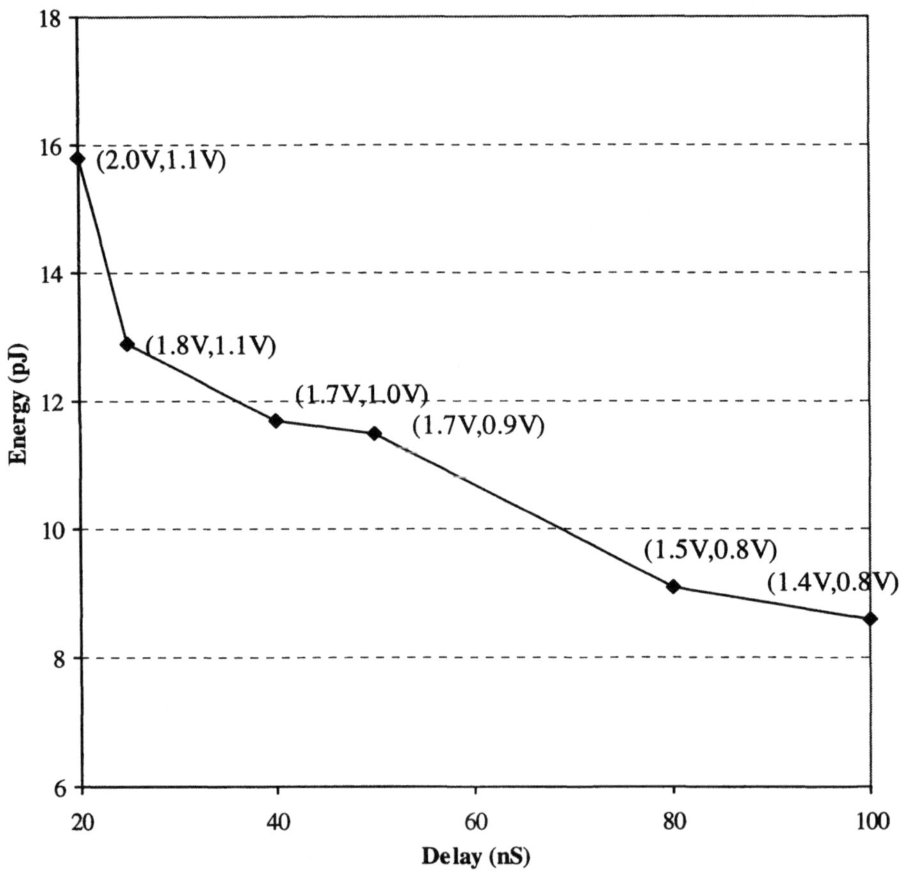

Figure 5. Energy - Delay Tradeoff

This exercise illustrates the robustness of the low-swing interconnect over a wide range of supply voltage. More importantly, this illustrates a method of dynamically balancing the energy consumption and speed performance. This can be exploited in systems that use dynamic voltage scaling to obtain the minimum energy while delivering the required performance [Burd01].

5 CONCLUSION

The final step in the validation process is the measurement of actual data from the physical implementation. Measured data are used to compare LP_PGAII with a commercial FPGA, XC4000XV. The effort in the redesign of the interconnect architecture pays off with ten to fifty times lower energy as compared to the commercial architecture. The execution energy is lower in the low-energy FPGA by a factor of five to sixteen.

The difference in the configuration overhead between the two FPGAs is quite significant, approximately four orders of magnitude. This is partly due to the different configuration techniques and partly due to the low energy techniques incorporated in the prototype. By reducing the transition activity during configuration, the energy associated with this step is reduced further by almost a factor of two.

To demonstrate the robustness of the low swing circuitry and to explore energy-delay tradeoff, measurements are taken over a wide voltage range. The speed performance of the design can be increased by almost four times while consuming 80% more energy.

Chapter 9

CONCLUSION

1 FPGA: THE EVOLUTION

The domain of FPGAs has undergone a dramatic revolution in the past decade. What started as a cheap alternative to Mask Programmable Gate Arrays (MPGA) in the mid-eighties has now progressed into a new model of computation. The first step in the process was the recognition of the re-programmability of the architecture. This made it possible to delay the programming of the FPGA to the last minute, and even change the programming if an error was detected in the implemented circuit. The next step was the adoption of the FPGA by the prototyping community. The fact that the architecture can be programmed over and over again to try out actual hardware implementations of the design ideas was a desired capability.

The next step has been the recognition of the limits of general-purpose processors. The "generality" of the processor constrains it to a fixed set of predefined instructions. Even though present day processors are bolstered with an array of accelerator units, the problem of a fixed instruction depth exists. In such a scenario, the reconfigurability of the FPGA, and the bit-level control offered by the architecture, open up the possibility of an accelerator unit that can be customized on an application basis. This aspect has generated interest in the research community, and among commercial ventures.

2 ENERGY EFFICIENCY

The FPGA industry kept pace with demand, and fueled new applications by providing larger and faster architectures. This was made possible in part by catching up with the rest of the IC industry in process technology, and in recent years even leading the industry in adopting cutting-edge processes. By virtue of the dominant market for FPGAs, the main performance criteria were logic capacity and speed performance. Power dissipation was a secondary concern. The industry depended on the reduced feature sizes and lower supply voltage accompanying each process generation to keep the power manageable.

Designers unfortunately accepted the high power dissipation as an unavoidable side effect of the programmability of the architecture. One domain where the characteristics of the FPGA can be utilized effectively is the portable computing domain. This domain has to deal with a number of different data streams with associated standards and computation. The configurable computing platforms can be quite effective in these portable devices. The main problem is that the prohibitive energy consumption of the present commercial FPGA will limit its use.

The reduction of feature size to the sub-0.1μm region will also force the power dissipation issue. Although the reduction in feature size, accompanied by reduced parasitic capacitance and reduced voltage, will lower the power dissipation per logic gate, the overall power dissipation will increase, because of the larger logic density and the higher operation frequencies of future integrated circuits.

3 THIS WORK

The main focus of this low-energy FPGA design work was to determine whether the large energy consumption of the FPGA was an unavoidable characteristic of the architecture. The circuit overhead that makes it possible to reconfigure the FPGA also increases the total energy. The only way to reduce the energy was by evaluating the existing FPGA architectures from an energy perspective, and to devise optimizations at the architecture and circuit level. This section summarizes this research work, and highlights new areas pertaining to energy efficiency that should be explored further.

3.1 Interconnect Architecture and FPGA Energy

The first part of the work looked at commercial FPGA architectures to analyze the energy components. Preliminary data showed that the dominant component was the interconnect, accounting for ~65% of the total energy. Measured data indicate that the diffusion capacitance of the routing switches was responsible for almost all of the interconnect energy, which in turn is due to the large size of the transistors used to realize the routing switches. This design decision was probably aimed at reducing the series resistance of the routing switches, to improve the delay of long routes. This indicated that the interconnect architecture has to redesigned to effect any significant improvement in the energy efficiency.

3.2 Architectural Exploration

The second part of the work looked at possible architectural solutions to improve the energy efficiency. This involved the development of a software environment to explore the impact of architectural decisions on the energy efficiency of the FPGA. The exploration environment spans the complete implementation flow from the synthesis of the function to the implementation of the function on the target architecture. Based on this flow, the logic block and the interconnect were redesigned from an energy perspective.

A logic block capable of implementing a 5-input random logic or a 2-bit arithmetic function without wasting logic resources results in minimizing the energy dissipated in the interconnect resources. The routes were broadly classified into three classes: neighbor-to-neighbor, intermediate length routes, and routes that span a significant fraction of the array. The interconnect architecture that was chosen consists of three distinct structures, with each level aimed at each specific class of connections.

One weakness is the quality of the implementation tools. The generality of the exploration tool makes it less efficient than architecture-specific tools. A preliminary comparison was done with VPR [Betz97b]. For a Symmetric Mesh structure, the cost of the implementations using the exploration tool was 5%-10% higher than VPR. This is acceptable considering the flexibility of the exploration environment.

Another factor that has to be considered is the possibility of bias in the exploration tools. The different aspects of the logic and interconnect structures were guided by empirical data gathered using the placement and routing tool. Bias can exist in the exploration environment that leads to specific architectural features. Sufficient care was taken to minimize this.

Routing resources contribute heavily to the total area, delay, and energy of the FPGA. In such a situation, it makes sense to have routing that is just sufficient for the purpose. The routing resources required are a function of the size of the array, while the array size is dependent on the application. The cost of the FPGA should be abstracted early in the development process, so that the designer can make decisions on the required FPGA resources. This flexibility can quite possibly improve the energy efficiency.

3.3 Circuit Techniques

The third part of the work implemented circuit-level improvisations based on the architectural redesign and the characteristics of the FPGA environment. The improvements in the interconnect architecture made it possible to reduce contribution of the routing fabric to the interconnect energy. Low-swing signaling circuits were evaluated to determine their viability in an FPGA environment. The implemented low-swing technique reduced the interconnect energy by a factor of two, as compared to conventional full-swing signaling. Double-edge triggering combined with low-swing distribution was used to reduce the clock distribution energy by a factor of three.

One factor that was not considered is the leakage current. In LP_PGAII, the leakage current was of the order of tens of microamperes at a power supply of 1.5V. As the process generations move into the sub-0.1μm regime, accompanied by lower threshold voltages, the sub-threshold leakage currents will become more important. One immediate technique is the use of multiple threshold processes. This can be employed to realize the critical path using the low-threshold devices, while implementing the rest of the design using high-threshold devices to reduce the leakage current. This method can already be seen in thc advanced processes offered by most semiconductor fabrication facilities.

3.4 Configuration Energy

The fourth part of the work looked at the implication of the configuration technique on the configuration energy. This becomes important as the FPGA is used in an embedded environment as a performance accelerator. The implementation of the configuration step in the current commercial architectures leaves much room for improvement. As a first step in this direction, a random access technique of programming was used to facilitate fast and energy-efficient programming. The configuration was encoded to reduce the number of bits required to program the FPGA. This helped tp reduce the energy for programming the FPGA. It was demonstrated how software methods can be used to order the sequence of programming to minimize transition activity, and therefore energy. These methods resulted in the reduction of the programming energy by a few orders of magnitude.

Another important component of the configuration energy is the transfer of configuration bits from the storage device to the FPGA. Other configuration loading techniques like time-multiplexing and run-length encoding of the configuration have to be explored to solve this problem.

The basic concept of time-multiplexing is that each programmable point in the FPGA has multiple memory storage elements. By switching the memory bit that controls the programmable element, it is possible to switch the function implemented on the FPGA on a cycle basis [Trimberger97]. The FPGA can be visualized as having multiple programming planes. Even though programming each plane can be potentially more expensive than programming a single-context FPGA, if there is temporal locality among the functions implemented, the cost of transferring the configuration can be minimized.

Configuration compression has been used in this work to eliminate redundant configuration bits. Another level of reduction can be obtained by recognizing that in datapath functions, the different bit-slices have identical configurations. This makes it possible to program multiple locations using a single write operation. This has been exploited to program the XC6200 FPGA [Hauck99a]. Another method is to use compression techniques such as run-length encoding to reduce the data transfers on the bus [Hauck99b]. These methods have to be explored while taking into consideration the increased complexity involved.

3.5 Implementation

In the final part of the work, the energy reduction techniques were implemented in actual silicon. Three prototype FPGAs were built. The first prototype, LP_PGAII, was an array of sixty-four logic blocks. The purpose of this chip was to verify the architectural and circuit techniques aimed at reducing the execution energy. The second prototype was an embedded version of LP_PGAII. The array was used as an accelerator in a digital signal processor for voice band processing. Data obtained from the embedded FPGA verified the applicability of an FPGA in an energy-sensitive platform. This implementation also brought into focus the overhead associated with frequent reconfiguration of the FPGA. The last prototype, LP_PGAII incorporated the improvements to reduce the configuration energy. Measured data from the prototypes demonstrate five times to twenty-two times improvement in execution energy over comparable commercial architectures.

4 LOOKING AHEAD

Prior academic research projects are slowly showing their presence in the commercial world. Considerable work must be done before FPGAs will be accepted completely by the design community in the new role.

4.1 Programmable Systems

The model of computation based on the programmability of FPGAs has taken off in the commercial world. This is reflected in the different flavors of FPGAs or programmability offered by different vendors. The approach in this area can be broadly divided into two camps: programmability in ASICs and dedicated functionality in FPGAs. This is illustrated in Fig. 1.

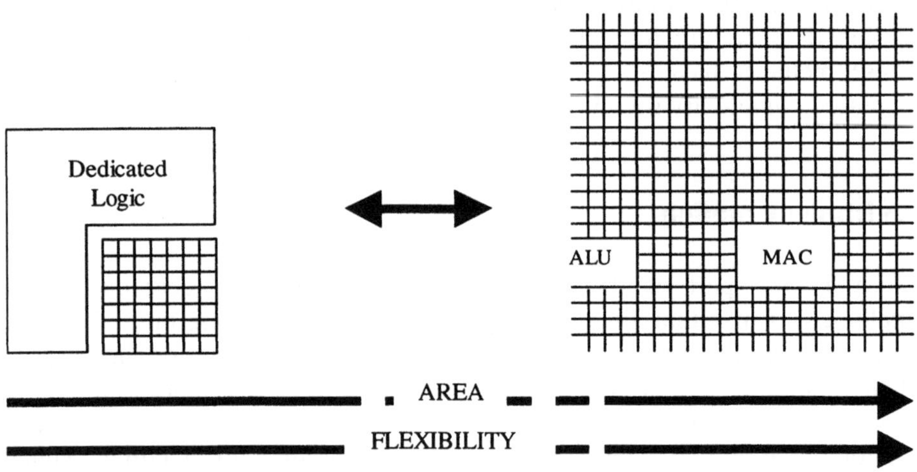

Figure 1. Programmable Systems

4.1.1 ASIC + FPGA

One approach is to use embedded FPGAs in a system dominated by dedicated logic. The purpose of the FPGA is to implement the functionality that is dynamic. This domain is very sensitive to the area of the implementation. The logic capacity of the embedded FPGA will most probably be constrained to tens of thousands of logic gates to control the area overhead.

Actel, with its Varicore series of FPGAs, is one of the commercial groups that supports this approach by supplying embedded FPGAs [Varicore].

4.1.2 FPGA + Dedicated Logic

The second approach is aimed at the traditional FPGA user who wants to pack more functionality in a given area. By embedding dedicated logic like multipliers and general-purpose processors, these functions can be instantiated in an application without paying the high area penalty if these were to be mapped to regular programmable blocks. Even though the area of the functional units is small, the bottleneck will be the routing resources required to connect to these complex blocks.

Almost all the major FPGA vendors now provide dedicated functional units embedded in their latest offerings. For example, the Virtex II [Xilinx00] devices from Xilinx will have PowerPC cores embedded in the array.

4.2 Design Flow

The dominant market for FPGA has been the stand-alone segment, and the software support has been tailored towards such a use of FPGAs. The design flow used in FPGAs has to keep up to support use of the FPGA in a wider range of applications as they become popular.

One of the main issues that has to be dealt with is a design flow that makes the use of these programmable systems efficient. Even though all the different architectures boast impressive performance gains, the deciding factor will be the software flow that can extract the claimed performance.

Consider the embedded FPGA in an ASIC system. One of the critical requirements is a seamless integration of the FPGA in an ASIC flow. The designer should be able to evaluate the cost of the FPGA and balance it with the achievable performance gains. This requires the abstraction of the FPGA at a high level so that the designer can accurately compare an FPGA implementation with other options. This is required before committing expensive silicon space to a specific array size of the FPGA. The compiler for the system should be able to efficiently schedule operations on the FPGA to justify the area cost.

4.3 Conclusion

FPGAs have traditionally been associated with poor energy and delay performance, with the blame being placed on the overhead of programmability. Preliminary analysis of existing FPGA architectures shows that the power consumption of the chips will only become worse as the process technology moves below the 0.1um region. This will severely limit the use of FPGAs in several energy-sensitive environments. The main purpose of this work had been to evaluate the FPGA architecture from an energy perspective, and decide if the high energy consumption is a penalty that has to be paid for the programmability. This work on the design of a low-energy FPGA has shown that with architectural and circuit design improvements the energy can be improved considerably.

The authors believe that this work is just a beginning in the direction of low-energy FPGA design. Depending on the application domain, specialization of the architecture will help in further reducing the energy.

BIBLIOGRAPHY

[74ALVC164245]
16-Bit Dual Supply Translating Transceiver, Philips Semiconductors.

[Abnous98]
A. Abnous, K. Seno, Y. Ichikawa, M. Wan, and J. Rabaey, "Evaluation of a Low-Power Reconfigurable DSP Architecture," *Proceedings Parallel and Distributed Processing. SPDP '98 Workshops*, Springer-Verlag, March 1998, pp. 55-60.

[Actel1]
ACT1 Series FPGAs, Actel Corporation, 1996.

[Actel2]
Accelerator Series FPGAs - ACT3 Family, Actel Corporation, 1997.

[Actel3]
SX Family of High Performance FPGAs, Actel Corporation, 2001.

[Actel4]
ProAsic 500K Family, Actel Corporation, 2000.

[Aggarwal94]
A. A. Aggarwal and D. M. Lewis, "Routing Architectures for Hierarchical Field Programmable Gate Arrays," *Proceedings IEEE International Conference on Computer Design: VLSI in Computers and Processors*, Cambridge, Massachusetts, 1994, pp. 475-478.

[Alexander94]
M. J. Alexander, J. P. Cohoon, J. L. Ganley, and G. Robins, "An Architecture-Independent Approach to FPGA Routing Based on Multi-weighted Graphs," *Proceedings EURO-DAC with EURO-VHDL*, New York, ACM, 1994, pp. 259-264.

[Alexander96]
M. J. Alexander and G. Robins, "New performance-driven FPGA routing algorithms," *IEEE Transactions on Computer-Aided Design of Integrated Circuits and Systems*, vol.15, no.12, December 1996, pp. 1505-1517.

[Altera1]
Private Communication.

[Altera2]
FLEX10K Embedded Programmable Logic Family Data Sheet, Altera, 2000.

[Altera3]
Classic EPLD Family Data Sheet, Altera, 2000.

[Altera97]
Flex 10K Power Consumption, Altera, Technical Brief 23, June 1997.

[Arnold93]
J. M. Arnold, D. A. Buell, D. T. Hoang, D. V. Pryor, N. Shirazi, and M. R. Thistle, "The Splash 2 Processor and Applications," *Proceedings IEEE International Conference on Computer Design: VLSI in Computers and Processors*, Massachusetts, October 1993, pp. 3-6.

[Bellman58]
R. Bellman, "On a Routing Problem," *Quarterly of Applied Mathematics*, 1958, pp. 87-90.

[Benes99]
M. Benes, *Design and Implementation of Communication and Switching Techniques for the Pleiades Family of Processors*, M.S. Thesis, University of California, Berkeley, December 1999.

[Bertin93]
P. Bertin, D. Roncin, and J. Vuillemin, "Programmable Active Memories: A Performance Assessment," *PRL Research Report #24*, 1993. Available online at http://research.compaq.com/PRL/ publications/PRL-PrlReport.html.

[Betz96]
V. Betz and J. Rose, "Directional Bias and Non-Uniformity in FPGA Global Routing Architectures," *IEEE/ACM International Conference on Computer-Aided Design*, San Jose, California, 1996, pp. 652–659.

[Betz97a]
V. Betz and J. Rose, "Cluster-Based Logic Blocks for FPGAs: Area-Efficiency vs. Input Sharing and Size," *IEEE Custom Integrated Circuits Conference*, Santa Clara, California, 1997, pp. 551–554.

[Betz97b]
V. Betz and J. Rose, "VPR: A New Packing, Placement and Routing Tool for FPGA Research," *Proceedings 7th International Workshop on Field-programmable Logic and Applications*, Berlin, Germany, 1997, pp. 213-22.

[Betz99a]
V. Betz and J. Rose, "FPGA Routing Architecture: Segmentation and Buffering to Optimize Speed and Density," *ACM/SIGDA International Symposium on Field Programmable Gate Arrays*, New York, 1999, pp. 59-68.

[Betz99b]
V. Betz and J. Rose, "Circuit Design, Transistor Sizing and Wire Layout of FPGA Interconnect," *Proceedings of the IEEE 1999 Custom Integrated Circuits Conference*, Piscataway, New Jersey, 1999, pp. 171-174.

[Borriello]
G Borriello, S. Hauck, and S Burns, "The Triptych FPGA Architecture," *IEEE Transactions on Very Large Scale Integration Systems*, vol. 3, no. 4, December, pp. 491-501.

[Brown92]
S. D. Brown, R. J. Francis, J. Rose and Z. G Vranesic, *Field-Programmable Gate Arrays*, Kluwer Academic Publishers, The Netherlands, 1992.

[Brown93]
S. D. Brown, J. Rose, and Z. G Vranesic, "A Stochastic Model to Predict the Routability of Field-Programmable Gate Arrays," *IEEE Transactions on Computer-Aided Design of Integrated Circuits and Systems*, vol. 12, no. 12, December 1993, pp. 1827-1838.

[Brown96]
S. Brown, G Lemieux, and M. Khellah, "Segmented Routing for Speed-Performance and Routability in Field-Programmable Gate Arrays," *Journal of VLSI Design*, 4(4), 1996, pp. 275-291.

[Burd00]
T. Burd, T. Pering, A. Stratakos, and R. Brodersen, "A Dynamic Voltage-scaled Microprocessor System," *IEEE Journal of Solid-State Circuits*, vol. 35, no. 11, November 2000, pp. 1571-1580.

[Chan96]
V. C. Chan and D. M. Lewis, "Area-Speed Tradeoffs for Hierarchical Field-Programmable Gate Arrays," *ACM Fourth International Symposium on Field-Programmable Gate Arrays*, New York, 1996, pp.51-57.

[Chow99a]
P. Chow, O. S. Soon, J. Rose, K. Chung, G Paez-Monzon, and I. Rahardja, "The Design of an SRAM-Based Field-Programmable Gate Array Architecture," *IEEE Transactions on Very Large Scale Integration (VLSI) Systems*, vol. 7, no. 2, June 1999, pp. 191-197.

[Chow99b]
P. Chow, O. S. Soon, J. Rose, K. Chung, G Paez-Monzon, and I. Rahardja, "The Design of an SRAM-Based Field-Programmable Gate Array-Part II: Circuit Design and Layout," *IEEE Transactions on Very Large Scale Integration (VLSI) Systems*, vol. 7, no. 3, September 1999, pp. 321-330.

[Chung91]
K. Chung, S. Singh, J. Rose, and P. Chow, "Using Hierarchical Logic Blocks to Improve the Speed of FPGAs," *International Workshop on Field Programmable Logic and Applications*, Oxford, UK, EE&CS Books, 1991, pp. 103-113.

[Colshan94]
R. Colshan and B. Jaroun, "A Novel Reduced Swing CMOS Bus Interface Circuit for High Speed Low Power VLSI Systems," *Proceedings of IEEE International Symposium on Circuits and Systems*, vol. 4, 1994, pp. 351-354.

[Cong94]
J. Cong and Y. Ding, "FlowMap: An Optimal Technology Mapping Algorithm for Delay Optimization in Lookup-Table Based FPGA Designs," *IEEE Transactions on CAD*, vol. 13, no. 1, Jan 1994, pp. 1-12.

[Cormen98]
T. H. Cormen, C. E. Leisserson, and R. L. Rivest, *Introduction to Algorithms*, The MIT Press, Cambridge, Massachusetts, 1990.

[Dally98]
W. J. Dally and J. W. Poulton, *Digital Systems Engineering*, Cambridge University Press, 1998.

[DeHon00]
A. DeHon, "The Density Advantage of Configurable Computing," *Computer*, vol. 33, no. 4, April 2000, pp. 41-49.

[Dijkstra59]
E. W. Dijkstra, "A Note on Two Problems in Connexion with Graphs," *Numerische Mathematik*, 1959, pp. 269-271.

[Ebcling95]
C. Ebeling, L. McMurchie, S. A. Hauck, and S. Burns. "Placement and Routing Tools for the Triptych FPGA," *IEEE Trnasactions on VLSI*, December 1995, pp. 473-482.

[Ebeling96]
C. Ebeling, D. C. Cronquist, and P. Franklin. "RaPiD - Reconfigurable Pipelined Datapath," *The 6th International Workshop on Field-Programmable Logic and Applications*, 1996, pp. 126-135.

[Elmore48]
E. Elmore, "The Transient Response of Damped Linear Networks with Particular Regard to Wideband Amplifiers," *Journal of Applied Physics*, January 1948, pp. 55-63.

[Ford62]
L. R. Ford and D. R. Fulkerson, *Flows in Networks*, Princeton University Press, 1962.

[Hamdy88]
E. Hamdy, J. McCollum, S. Chen, S. Chiang, S. Eltoukhy, J. Chang, T. Speers, and A. Mohsen, "Dielectric Based Antifuse for Logic and Memory ICs," *International Electron Devices Meeting, Technical Digest,* 1988, pp. 786-789.

[Hauck99a]
S. Hauck and W. D. Wilson, "Runlength Compression Techniques for FPGA Configurations," *Seventh Annual IEEE Symposium on Field-Programmable Custom Computing Machines*, Los Alamitos, California, 1999, pp. 286-287.

[Hauck99b]
S. Hauck, L. Zhiyuan, and E. Schwabe, "Configuration Compression for the Xilinx XC6200 FPGA," *IEEE Transactions on Computer-Aided Design of Integrated Circuits and Systems*, vol. 18, no. 8, IEEE, August 1999, pp. 1107-1113.

[He93]
J. He and J. Rose, "Advantages of Heterogeneous Logic Block Architecture for FPGAs," *Proceedings of the IEEE Custom Integrated Circuits Conference*, San Diego, California, 1993, pp. 7.4.1-7.4.5.

[Hiraki95]
M. Hiraki, H. Kojima, H. Misawa, T. Akazawa, and Y. Hatano, "Data-Dependent Logic Swing Internal Bus Architecture for Ultralow-Power LSI's," *IEEE Journal of Solid-State Circuits*, vol. 30, no. 4, April 1995, pp. 397-402.

[Hoang93]
D. T. Hoang, "Searching Genetic Databases on Splash 2," *Proceedings IEEE Workshop on FPGAs for Custom Computing Machines*, Napa, California, April 1993, pp. 5-7.

[Hwang92]
F. K. Hwang, D. S. Richards, and P. Winter, *The Steiner Tree Problem*, North-Holland, 1992.

[Ingber93]
L. Ingber, "Simulated Annealing: Practice Versus Theory," *Journal of Mathematical Computer Modeling*, vol. 18, no. 11, December 1993, pp. 29-57.

[ITRS99]
International Technology Roadmap for Semiconductors, Semiconductor Industry Association, 1999.

[Kahng95]
A. B. Kahng and G. Robins, *On Optimal Interconnections for VLSI*, Kluwer Academic Publishers, The Netherlands, 1995.

[Kirkpatrick83]
S. Kirkpatrick, C. Gelatt Jr., and M. P. Vecchi, "Optimization by Simulated Annealing," *Science*, vol. 220, 1983, pp. 671-680.

[Kou81]
L. Kou, G. Markowsky, and L. Berman, "A Fast Algorithm for Steiner Trees," *Acta Informatica*, 15(1981), pp. 141-145.

[Kouloheris91]
J. L. Kouloheris and A. El Gamal, "FPGA Performance Versus Cell Granularity," *Proceedings of the IEEE Custom Integrated Circuits Conference*, San Diego, California, 1991, pp. 6.2.1-6.2.4.

[Kouloheris92]

J. L. Kouloheris and A. El Gamal, "PLA-Based FPGA Area Versus Cell Granularity," *Proceedings of the IEEE Custom Integrated Circuits Conference*, Boston, Massachusetts, 1992, pp. 4.3.1-4.3.4.

[Kusse97]

E. Kusse, *Analysis and Circuit Design for Low Power Programmable Logic Modules*, M.S. Thesis, University of California, Berkeley, December 1997.

[Lai97]

Y. Lai and P. Wang, "Hierarchical Interconnection Structures for Field Programmable Gate Arrays," *IEEE Transactions on Very Large Scale Integration Systems*, vol.5, no.2, June 1997, pp.186-196.

[Lai98]

Y. Lai, C. Kao, T. Chang, and K. Chen, "A Field Programmable Gate Array Chip with Hierarchical Interconnection Structure," *Proceedings of the 1998 IEEE International Symposium on Circuits and Systems*, Monterey, California, 1998, pp. 402-405.

[Lam88]

J. Lam and J. M. Delosme, "Performance of a New Annealing Schedule," *Proceedings 25th ACM/IEEE Design Automation Conference*, New York, 1988, pp. 306-311.

[Lemieux93]

G. Lemieux and S. Brown, "A Detailed Router for Allocating Wire Segments in FPGAs," *ACM Physical Design Workshop*, Lake Arrowhead, California, April 1993, pp. 215-226.

[Liu94]

D. D. Liu and C. Svensson, "Power Consumption Estimation in CMOS VLSI Chips," *IEEE Journal of Solid-State Circuits*, vol. 29, no. 6, June 1994, pp. 663-670.

[Llopis96]

R. P. Llopis and M. Sachdev, "Low Power, Testable Dual Edge Triggered Flip-Flops," *Proceedings 1996, International Symposium on Low Power Electronics and Design*, Monterey, California, August 1996, pp.341-345.

[Mirsky96]

E. Mirsky and A. DeHon, "MATRIX: A Reconfigurable Computing Architecture with Configurable Instruction Distribution and Deployable Resources," *Proceedings IEEE Symposium on FPGAs for Custom Computing Machines*, Napa, California, April 1996, pp. 17-19.

[Murgai90]

R. Murgai, Y. Nishizaki, N. Shenoy, R. K. Brayton, and A. Sangiovanni-Vincentelli, "Logic Synthesis for Programmable Gate Arrays," *Proceedings ACM/IEEE Design Automation Conference*, Orlando, Florida, June 1990, pp. 24-28.

[Nakagome93]

Y. Nakagome, K. Itoh, M. Isoda, K. Takeuchi, and M. Aoki, "Sub-1-V Swing Internal Bus Architecture for Future Low-Power ULSIs," *IEEE Journal of Solid-State Circuits*, vol. 28, no. 4, April 1993, pp. 414-419.

[Ochotta98]

E. S. Ochotta, P. J. Crotty, C. R. Erickson, C. T. Huang, R. Jayaraman, R. C. Li, J. D. Linoff, L. Ngo, H. V. Nguyen, K. M. Pierce, D. P. Wieland, J. Zhuang, and S. S. Nance, "A Novel Predictable Segmented FPGA Routing Architecture," *ACM/SIGDA*

International Symposium on Field Programmable Gate Arrays, New York, 1998, pp. 3-11.

[Rose90a]
J. Rose, R. J. Francis, D. Lewis, and P. Chow, "Architecture of Field-Programmable Array: The Effect of Logic Block Functionality on Area Efficiency," *IEEE Journal of Solid State Circuits*, vol. 25, no. 5, October 1990, pp. 1217-1225.

[Rose90b]
J. Rose and S. Brown, "The Effect of Switch Box Flexibility on Routability of Field-Programmable Gate Arrays," *Procedings 1990 Custom Integrated Circuits Conference*, May 1990, pp. 27.5.1-27.5.4.

[Rose97]
J. Rose and S. Brown, "Flexibility of Interconnection Structures for Field Programmable Gate Array," *IEEE Journal of Solid State Circuits*, vol. 26, no. 3, March 1997, pp. 277-282.

[Sechen85]
C. Sechen and A. Sangiovanni-Vincentelli, "The Timberwolf Placement and Routing Package," *Journal of Solid-State Circuits*, vol. SC-20, no. 2, April 1985, pp. 510-522.

[Singh92]
S. Singh, J. Rose, P. Chow, and D. Lewis, "The Effect of Logic Block Architecture on FPGA Performance," *IEEE Journal of Solid-State Circuits*, vol. 27, no. 3, March 1992, pp. 281-287.

[Swartz90]
W. Swartz and C. Sechen, "New Algorithms for the Placement and Routing of Macro Cells," *IEEE International Conference on Computer-Aided Design*, IEEE Computer Society Press, Washington, 1990, pp. 336-339.

[Trimberger94]
S. M. Trimberger, *Field-Programmable Gate Array Technology*, Kluwer Academic Publishers, The Netherlands, 1994.

[Trimberger97]
S. Trimberger, D. Carberry, A. Johnson, and J. Wong, "A Time-Multiplexed FPGA," *Proceedings The 5th Annual IEEE Symposium on Field-Programmable Custom Computing Machines*, Los Alamitos, California, 1997, pp. 22-28.

[Tsu99]
W. Tsu, K. Macy, A. Joshi, R. Huang, N. Walker, T. Tung, O. Rowhani, V. George, J. Wawrzynek, and A. DeHon, "HSRA: High-Speed, Hierarchical Synchronous Reconfigurable Array," *ACM/SIGDA International Symposium on Field Programmable Gate Arrays*, New York, 1999, pp.125-134.

[Tsutsui98]
A. Tsutsui and T. Miyazaki, "ANT-on-YARDS: FPGA/MPU Hybrid Architecture for Telecommunication Data Processing," *IEEE Transactions on Very Large Scale Integration (VLSI) Systems*, vol.6, June 1998, pp. 199-211.

[Varicore]
VariCore, Actel, 2001.

[Wazlowki93]
M. Wazlowski, "PRISM –II Compiler and Architecture," *Proceedings IEEE Workshop on FPGAs for Custom Computing Machines*, Napa, California, April 1993, pp. 5-7.

[Wittig96]
R. D. Wittig and P. Chow, "OneChip: An FPGA Processor with Reconfigurable Logic," *Proceedings IEEE Symposium on FPGAs for Custom Computing Machines*, April 1996, pp.126-135.

[XC4000XV]
XC4000XV FPGAs, Xilinx. Online at http://www.xilinx.com/products/xc4000xv.htm.

[XESS]
Xilinx Student Edition with Foundation 2.1 Software and XS40-005XL FPGA Prototyping Board, XESS Corporation, North Carolina.

[Xilinx00]
Platform FPGA- The Future of Logic Design, *Xcell-The Quarterly Journal for Xilinx Programmable Logic Users*, Fourth Quarter, 2000.

[Xilinx1]
Private Communication.

[Xilinx2]
The Programmable Logic Data Book, Xilinx, San Jose, 1998.

[Xilinx97]
B. Fawcett, "FPGAs, Power and Packages," *Xcell-The Quarterly Journal for Xilinx Programmable Logic Users*, Second Quarter, 1997, pp. 2-4.

[Xilinx98]
XC4000XL Power Calculation, *Xcell-The Quarterly Journal for Xilinx Programmable Logic Users*, First Quarter, 1998.

[Yamauchi95]
H. Yamauchi, H. Akamatsu, and T. Fujita, "An Asymptotically Zero Power Charge-Recycling Bus Architecture for Battery-Operated Ultrahigh Data Rate ULSI's," *IEEE Journal of Solid-State Circuits*, vol. 30, no. 4, April 1995, pp. 423-431.

[Zhang98]
H. Zhang and J. Rabaey, "Low-Swing Interconnect Interface Circuits," *Proceedings 1998 International Symposium on Low Power Electronics and Design*, Monterey, California, August 1998, pp. 161-166.

[Zhang00a]
H. Zhang, V. Prabhu, V. George, M. Wan, M. Benes, A. Abnous, and J. Rabaey, "A 1 V Heterogeneous Reconfigurable Processor IC for Baseband Wireless Applications," *International Solid State Circuits Conference*, February 2000, pp. 68-69.

[Zhang00b]
H. Zhang, V. George, and J. Rabaey, "Low-swing On-chip Signaling Techniques: Effectiveness and Robustness," *IEEE Transactions on VLSI Systems*, vol. 8, no. 3, June 2000, pp. 264-27.

INDEX

V

MIX
Papier aus verantwortungsvollen Quellen
Paper from responsible sources
FSC® C105338

If you have any concerns about our products,
you can contact us on
ProductSafety@springernature.com

In case Publisher is established outside the EU,
the EU authorized representative is:
Springer Nature Customer Service Center GmbH
Europaplatz 3, 69115 Heidelberg, Germany

Printed by Libri Plureos GmbH
in Hamburg, Germany